ELECTRODEPOSITION

ELECTRODEPOSITION

The Materials Science
of Coatings and Substrates

by

Jack W. Dini

Lawrence Livermore National Laboratory
Livermore, California

Reprint Edition

NOYES PUBLICATIONS
WILLIAM ANDREW PUBLISHING, LLC

Norwich, New York, U.S.A.

Copyright © 1993 by Jack W. Dini

No part of this book may be reproduced or utilized in
any form or by any means, electronic or mechanical,
including photocopying, recording or by any information storage and retrieval system, without permission
in writing from the Publisher.

Library of Congress Catalog Card Number: 92-27804
ISBN-13: 978-0-8155-1320-9   ISBN-10: 0-8155-1320-8

Transferred to Digital Printing 2012

Published in the United States of America by
Noyes Publications/William Andrew Publishing, LLC
13 Eaton Avenue, Norwich, New York 13815

Library of Congress Cataloging-in-Publication Data

Dini, J.W.
    Electrodeposition : the materials science of coatings and
substrates / by Jack W. Dini.
        p.   cm.
    Includes bibliographical references and index.
    ISBN-13: 978-0-8155-1320-9   ISBN-10: 0-8155-1320-8
    1. Electroplating. I. Title.
TS670.D55   1992
671.7'32--dc20   92-27804
                                          CIP

MATERIALS SCIENCE AND PROCESS TECHNOLOGY SERIES

Editors

Rointan F. Bunshah, University of California, Los Angeles *(Series Editor)*
Gary E. McGuire, Microelectronics Center of North Carolina *(Series Editor)*
Stephen M. Rossnagel, IBM Thomas J. Watson Research Center *(Consulting Editor)*

Electronic Materials and Process Technology

(continued)

HANDBOOK OF CHEMICAL VAPOR DEPOSITION: by Hugh O. Pierson

DIAMOND FILMS AND COATINGS: edited by Robert F. Davis

ELECTRODEPOSITION: by Jack W. Dini

Ceramic and Other Materials—Processing and Technology

SOL–GEL TECHNOLOGY FOR THIN FILMS, FIBERS, PREFORMS, ELECTRONICS AND SPECIALTY SHAPES: edited by Lisa C. Klein

FIBER REINFORCED CERAMIC COMPOSITES: by K.S. Mazdiyasni

ADVANCED CERAMIC PROCESSING AND TECHNOLOGY—Volume 1: edited by Jon G.P. Binner

FRICTION AND WEAR TRANSITIONS OF MATERIALS: by Peter J. Blau

SHOCK WAVES FOR INDUSTRIAL APPLICATIONS: edited by Lawrence E. Murr

SPECIAL MELTING AND PROCESSING TECHNOLOGIES: edited by G.K. Bhat

CORROSION OF GLASS, CERAMICS AND CERAMIC SUPERCONDUCTORS: edited by David E. Clark and Bruce K. Zoitos

HANDBOOK OF INDUSTRIAL REFRACTORIES TECHNOLOGY: by Stephen C. Carniglia and Gordon L. Barna

CERAMIC FILMS AND COATINGS: edited by John B. Wachtman and Richard A. Haber

Related Titles

ADHESIVES TECHNOLOGY HANDBOOK: by Arthur H. Landrock

HANDBOOK OF THERMOSET PLASTICS: edited by Sidney H. Goodman

SURFACE PREPARATION TECHNIQUES FOR ADHESIVE BONDING: by Raymond F. Wegman

FORMULATING PLASTICS AND ELASTOMERS BY COMPUTER: by Ralph D. Hermansen

HANDBOOK OF ADHESIVE BONDED STRUCTURAL REPAIR: by Raymond F. Wegman and Thomas R. Tullos

CARBON–CARBON MATERIALS AND COMPOSITES: edited by John D. Buckley and Dan D. Edie

PREFACE

INTRODUCTION

This book grew out of a short course that I've taught with Mort Schwartz and then Dick Baker. When Rudy Johnson (my colleague for many years at Sandia) saw my notes he suggested that, they should be put in the form of a book. His suggestion was the prod that I needed to undertake this project.

Knowledge of the relationships between microstructure and mechanical properties has always been one of the primary goals of metallurgy and materials science. This has led to the point that "tailored" properties of materials are now a reality. We are no longer limited to combinations of elements allowed by nature to form structures dictated by the small differences in the forces between atoms and the free energies in condensed matter that define equilibrium form.

The fascinating field of electrodeposition allows one to "tailor" the surface properties of a bulk material or in the case of electroforming, the entire part. Deposits can be produced to meet a variety of demands of the designer. For this reason and for the possibilities that exist in terms of "new materials" for a variety of applications, a thorough understanding of materials science and principles is of utmost importance. This book is intended to provide some of that understanding.

The sequence of chapters in the book takes the reader from the substrate to the outer surface of the coating. It starts with the substrate (Hydrogen Embrittlement), then proceeds to the substrate/coating interface (Adhesion, Diffusion), then the bulk of the coating (Structure, Properties, Addition Agents, Stress, Porosity) and finally to the environmental interface of the coating (Corrosion, Wear).

Although the title *ELECTRODEPOSITION: The Materials Science of Coatings and Substrates* sounds specific to electrodeposition, this does not mean that other coating processes are not covered. Coating technologies such as physical vapor deposition, chemical vapor deposition, plasma spraying and ion implantation are occasionally discussed. Growth of films and their performance from a materials science viewpoint have a great deal of commonality regardless of the technique used to deposit the film.

I am indebted to many individuals who have provided help along the way. Ted Pavlish opened my eyes to the fascinating field of electroplating in 1950 and he and his brother Arnold nurtured me for 9 years at Cleveland Supply Company (now Pavco) during my high school and college years. Following this, a few years at Battelle Columbus Laboratories gave me the chance to work with people like Glenn Schaer, Bill Safranek, John Beach, Charlie Faust, and Hugh Miller, some of the forerunners in electroplating research and development at that time. At Sandia Livermore, Rudy Johnson and I were a team for over 15 years and even though we've been apart for the past 13 years, some people still mistake one of us for the other. It was always difficult to determine where Rudy's effort stopped and mine began and vice versa. Also at Sandia, there were a number of highly talented, well educated young materials scientists and engineers who were quite helpful. Sandia's practice of hiring top-of-the-class people from the leading schools in the country meant that you learned by just being around such talent. My thanks to all of these people and others who were so helpful during the years at Sandia. At LLNL, a key individual was Harold Wiesner who was always available for advice and suggestions. Others at LLNL who have been particularly helpful include Ron Reno, Fritz Rittmann, Den Fisher and Tom Beat. Special thanks are due to Mort Schwartz for his support in introducing me to and sharing short course assignments as well as results of his efforts with graduate students at UCLA. I also thank all of you have helped either through AESF or on mutual technical projects. Lastly, but not least, thanks to George Narita of Noyes Publications for his help and encouragement in seeing this book to fruition.

Livermore, California   J. W. Dini
April, 1992

NOTICE

CONTENTS

1
INTRODUCTION

Materials science is the scientific discipline that probes the relations among structure, composition, synthesis, processing, properties, and performance in material systems. Eight major U.S. industries rely heavily on materials science: aerospace, automobile manufacturing, biomaterials, chemicals, electronics, energy, metals and telecommunications. Together these industries employ seven million people and account, for sales of $1.4 trillion (1)(2). A recent report of the National Research Council notes that advances have come in all eight industries from improved instrumentation, better controls on composition of products and expanded use of computers in modeling behavior of materials. According to the report: "The field of materials science and engineering is entering a period of unprecedented intellectual challenge and productivity. Scientists and engineers have a growing ability to tailor materials from the atomic scale upward to achieve desired functional properties" (1)(3). The field of materials science has grown to a major and distinct field since its origin in the 1940's. It is advancing at a revolutionary pace and is now generally recognized as being among the key emerging technological fields propelling our world society into the twenty-first century (4).

There are excellent texts on the general topic of materials science and a comprehensive 6000 page source of information consisting of eight volumes containing 1580 materials science topics(5). Originally published in 1986, this encyclopedia set has already been supplemented with two additional volumes (113 topics, 653 pages, in 1988, see reference 6; and 130 topics, 832 pages, in 1990, see reference 7) attesting to the continued expansion of materials science.

Since the purpose of this book is to relate materials science and

1

electrodeposition some brief history is in order. The 1949 text by Blum and Hogaboom (8) presents some of the principles of materials science even though it was not so named at that time. The book contains a number of photographs of structures of electrodeposits and data on properties. In the 1960's Read extended this coverage by showing the remarkable range of structures and properties that can be achieved by electrodepositing a given metal in a variety of ways (9,10). In more recent times (1982 and 1984) Weil introduced the topic of materials science of electrodeposits disclosing how the principles of materials science can be used to explain various structures of electrodeposits and how these structures influence properties (11,12). As Weil stated: "The understanding that has been gained is to a great extent responsible for changing plating from an art to a science" (11). Safranek's treatises on properties of deposits (1974 and 1986) are also very valuable resources (13)(14). These two volumes contain property data from over 1000 technical papers.

COMMENTS ON ELECTRODEPOSITION

Electrodeposition is an extremely important technology. Covering inexpensive and widely available base materials with plated layers of different metals with superior properties extends their use to applications which otherwise would have been prohibitively expensive (15). However, it should be noted that electroplating is not a simple dip and dunk process. It is probably one of the most complex unit operations known because of the unusually large number of critical elementary phenomena or process steps which control the overall process (16). An excellent example is the system model from Rudzki (Figure 1) for metal distribution showing the interrelation of plating variables and their complexity (17). Figure 2 is a simplified version summarizing the factors that influence the properties of deposits. Electrodeposition involves surface phenomena, solid state processes, and processes occurring in the liquid state, thereby drawing on many scientific disciplines as shown in Table 1 (15).

FACTORS AFFECTING COATINGS

It has been suggested that three different zones; 1) the substrate interface, 2) the coating, and 3) the coating-environment interface have to be considered when protecting materials with coatings (18). These, plus a fourth zone- the substrate, are covered in sequential fashion in the following chapters. Figure 3 shows these zones along with the titles of the chapters.

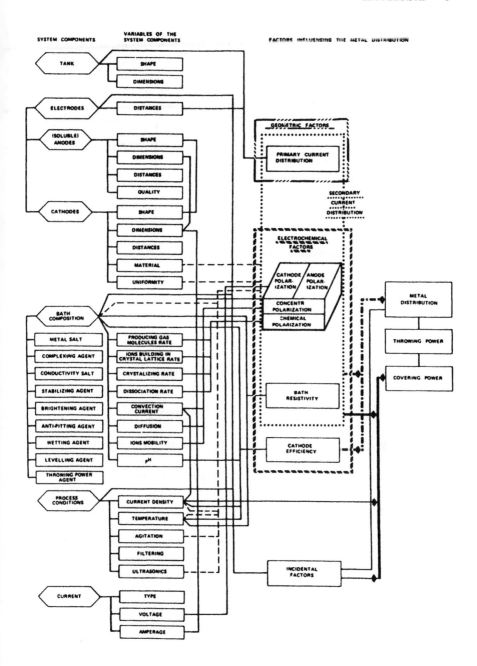

Figure 1: System model illustrating metal distribution relationships. From Reference 17. Reprinted with permission of ASM International, Metals Park, Ohio.

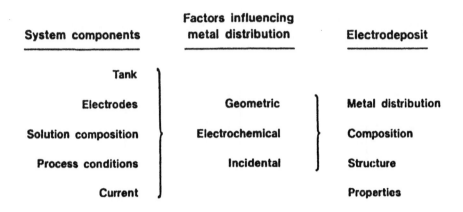

Figure 2: Metal distribution relationships in electrodeposition.

Table 1: Interdisciplinary nature of Electrodeposition*

Discipline	Involvement
Electrochemistry	Electrode processes
Electrochemical engineering	Transport phenomena
Surface science	Analytical tools
Solid state physics	Use of quantam mechanical solid state concepts to study electrode processes
Metallurgy and materials science	Properties of deposits
Electronics	Modern instrumentation

* From Reference 15.

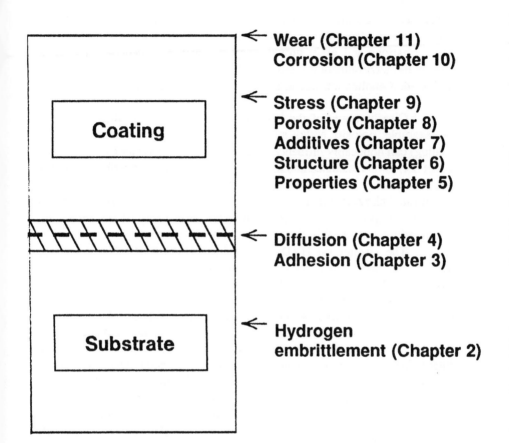

Figure 3: Important criteria when selecting coatings. This also is a listing of the following chapters in this book starting with HYDROGEN EMBRIT-TLEMENT and proceeding through to WEAR.

First is the substrate where potential hydrogen embrittlement effects are of concern. The second zone is the basis metal interface where adhesion of the coating and interdiffusion between the coating and substrate are of importance. The third zone is the coating itself where composition and microstructure determine properties and factors such as stress, phase transformations and grain growth exert noticeable influences. The final zone is the environmental interface where the interaction of the coating in its intended application has to be considered in terms of corrosion and/or wear.

Clearly, many of the items are important in more than one zone. For example; porosity and/or stress in the substrate (rather than just in the coating) can noticeably influence coating properties; porosity can noticeably affect corrosion resistance and tensile properties; hydrogen embrittlement is a factor not only for substrates but also for some coatings; and diffusion of codeposited alloying impurities to the surface can noticeably affect wear and corrosion properties. For reasons such as these, many of the topics are discussed interchangeably throughout the book.

The chapter on HYDROGEN EMBRITTLEMENT concentrates heavily on steels since these substrates are particularly susceptible to damage by hydrogen. This chapter also covers permeation of hydrogen through various protective coatings, hydrogen embrittlement of electroless copper deposits and hydrogen concerns as a result of chemical milling.

The importance of ADHESION is discussed in the next chapter and this topic is broken down into four categories; interfacial adhesion, interdiffusion adhesion, intermediate layer adhesion and mechanical interlocking. A variety of quantitative tests for measuring adhesion are discussed and then a methodology is presented for use when confronted with difficult-to-plate substrates. Processes that have been used to provide adhesion of coatings on difficult-to-plate substrates are discussed and supported with quantitative data. The relatively new approach of combining physical vapor deposition with electroplating which offers considerable promise for obtaining adherent bonds between coatings and difficult-to-plate substrates is also covered. Other techniques such as interface tailoring, alloying surface layers with metals exhibiting a high negative free energy of formation, use of partial pressure of various gases during deposition, reactive ion mixing and phase-in deposition are also discussed.

DIFFUSION, which is the attempt of a system to achieve equilibrium through elimination of concentration gradients, can result in degradation of properties and appearance. Diffusion mechanisms are discussed with particular emphasis placed on Kirkendall voids which can lead to loss of adhesion. Diffusion is influenced by the nature of the atoms, temperature, concentration gradients, nature of the lattice crystal structure, grain size, amount of impurities and the presence of cold work (19). An effective way to minimize or eliminate potential diffusion problems is the use of barrier coatings. In some instances, diffusion can benefit coating applications. Examples include deposition of alloy coatings and diffusion welding which utilizes diffusion to produce high integrity joints in a range of both similar and dissimilar metals.

Although PROPERTIES are discussed throughout the book, this chapter is included to cover some specifics not covered elsewhere. Topics include tensile property measurements, strength and ductility of thin deposits, the Hall-Petch relationship between strength and grain size, and the influence of impurities on properties. Superplasticity, which refers to the ability of a material to be stretched to many times its original length, is covered since electrodeposition offers some potential in this area.

STRUCTURE is one of the longest chapters in the book and rightfully so since structure is so dominant a factor in materials science. The variety of structures obtainable with electrodeposits are discussed and illustrated, as is the influence of substrate on coating structure. Phase transformations, which can noticeably affect deposit properties, are reviewed for electroless nickel, gold-copper, tin-nickel, palladium and cobalt deposits. Microstructural stability of copper and silver deposits at room temperature is also covered. Texture of deposits is an important structural parameter for bulk materials and coatings and a good illustration of how properties can be tailored for applications such as formability, corrosion resistance, etching characteristics, contact resistance, magnetics, wear resistance, and porosity. Fractals, which offer the materials scientist a new way to analyze micro-structures, provide a new tool for studying surfaces and corrosion processes. Some concepts of fractals are presented as are examples of results already obtained with various surfaces.

ADDITIVES are included as a separate chapter because of their extreme importance on the structure and properties of deposits. Some of the folklore regarding addition agents is discussed and examples included illustrating the interesting history of this complex aspect of electrodeposi-tion. Data are presented showing the influence of additives on tensile properties, leveling, and brightening. Typical additive systems used for deposition of a variety of deposits are reviewed as are proposed mechanisms of additive behavior. Control of addition agents via techniques such as the Hull cell, bent cathode, electroanalytical techniques, chromatography and other analytical methods is covered in some detail.

POROSITY is one of the main sources of discontinuities in electrodeposited coatings. It can noticeably influence corrosion resistance, mechanical properties, electrical properties and diffusion characteristics. Items which influence porosity include the substrate, the plating solution and its operating characteristics, and post plating treatments. An effective way to minimize porosity is to use an underplate. Another is to deposit coatings with specific crystallographic orientations which can strongly influence covering power and rate of pore closure. A variety of porosity tests are available for testing coatings and these are discussed in some detail.

STRESS in coatings can also adversely affect properties. A variety of options are available for reducing deposit stress and these include: choice of substrate, choice of plating solution, use of additives and use of higher plating temperatures. A variety of theories have been postulated regarding the origins of stress but none of them covers all situations. Numerous stress measurement techniques are available and they vary from the simple rigid strip technique to sophisticated methods using holographic interferometry.

CORROSION is affected by a variety of factors including metallurgical, electrochemical, physical chemistry and thermodynamic. Since all of these encompass the field of materials science, the topic of corrosion is essentially covered in many places in this book other than this particular chapter. Often it is difficult to separate corrosion from many of the other property issues associated with deposits. When selecting a coating it is important to know its position with respect to its substrate in the galvanic series for the intended application. Besides galvanic effects, the substrate and the interfacial zone between it and the coating can noticeably affect the growth and corrosion resistance of the subsequent coating since corrosion is affected by structure, grain size, porosity, metallic impurity content, interactions involving metallic underplates and cleanliness or freedom from processing contaminants (20). Decorative nickel-chromium coatings developed for automotive industry applications are a good example of use of materials science and electrochemistry to improve corrosion resistance properties.

WEAR, like corrosion, does not fit handily within the confines of a traditional discipline. Physics, chemistry, metallurgy and mechanical engineering all contribute to this topic. A particular feature of electrodeposition that is attractive for wear applications is its low temperature processing and ability to be applied to distortion prone substrates without increasing stress in the composite. Mechanisms of wear are discussed as are some of the more important tests used for evaluating wear characteristics. Coatings that are used for various wear applications include chromium, electroless nickel, precious metals and anodized aluminum. Recent advances include ion implantation of chromium deposits with nitrogen to provide improved wear resistance and codeposition of dispersed particles with electroless nickel. Microlayered metallic coatings also known as composition modulated coatings also offer promise. Although electrodeposited coatings are typically not effective at temperatures above 500°C, composite coatings containing chromium or cobalt particulates in a nickel or cobalt matrix can be effective.

REFERENCES

1. *Materials Science and Engineering for the 1990's-Maintaining Competitiveness in the Age of Materials*, By the Committee on Materials Science and Engineering of the National Research Council, P. Chaudhari and M. Flemings, Chairmen, National Academy Press, Washington, D.C. (1989).

2. R. Abbaschian, "Materials Education-A Challenge", *MRS Bulletin*, Vol XV, No 8, 18 (Aug 1990).

3. P. H. Abelson, "Support for Materials Science and Engineering", *Science*, 247, 1273 (16 March 1990).

4. A. L. Bement, Jr., "The Greening of Materials Science and Engineering", *Metallurgical Transactions A*, 18A, 363 (March 1987).

5. *Encyclopedia of Materials Science and Engineering*, M. B. Bever, Editor-in-Chief, Pergamon Press, Oxford, (1986).

6. *Encyclopedia of Materials Science and Engineering*, Supplementary Volume 1, R. W. Cahn, Editor, Pergamon Press, Oxford, (1988).

7. *Encyclopedia of Materials Science and Engineering*, Supplementary Volume 2, R. W. Cahn, Editor, Pergamon Press, Oxford, (1990).

8. W. Blum and G. B. Hogaboom, *Principles of Electroplating and Electroforming*, Third Edition, McGraw-Hill (1949).

9. H. J. Read, "The Effects of Addition Agents on Physical and Mechanical Properties of Electrodeposits", *Plating*, 49, 602 (1962).

10. H. J. Read, "Metallurgical Aspects of Electrodeposits", *Plating* 54, 33 (1967).

11. R. Weil, "Materials Science of Electrodeposits", *Plating & Surface Finishing*, 69, 46 (Dec 1982).

12.   R. Weil, "Relevant Materials Science for Electroplaters", *Electroplating Engineering Handbook*, Fourth Edition, L. J. Durney, Editor, Van Nostrand Reinhold, (1989).

13.   W. H. Safranek, *The Properties of Electrodeposited Metals and Alloys-A Handbook*, American Elsevier Publishing Co. (1974).

14.   W. H. Safranek, *The Properties of Electrodeposited Metals and Alloys, A Handbook, Second Edition*, American Electroplaters & Surface Finishers Soc., (1986).

15.   U. Landau, "Plating-New Prospects for an Old Art", *Electrochemistry in Industry, New Directions*, U. Landau, E. Yeager and D. Kortan, Editors, Plenum Press, New York (1982).

16.   V. A. Ettel, "Fundamentals, Practice and Control in Electrodeposition-An Overview", *Application of Polarization Measurements in the Control of Metal Deposition*, I. H. Warren, Editor, Elsevier, Amsterdam (1984).

17.   G. J. Rudzki, *Surface Finishing Systems*, ASM, Metals Park, Ohio (1983).

18.   H. Holleck, "Material Selection for Hard Coatings", *J. Vac. Sci. Technol.*, A4(6), 2661 (Nov/Dec 1986).

19.   "Diffusion and Surface Treatments", Chapter 13 in *Elements of Metallurgy*, R. S. Edelman, Editor, ASM, Metals Park, Ohio (1963).

20.   R. P. Frankenthal, "Corrosion in Electronic Applications", Chapter 9 in *Properties of Electrodeposits, Their Measurement and Significance*, R. Sard, H. Leidheiser, Jr., and F. Ogburn, Editors, The Electrochemical Soc. (1975).

2

HYDROGEN EMBRITTLEMENT

Electrodeposition and electroless deposition and their associated processing steps including acid pickling and electrocleaning can generate hydrogen which can enter substrates in the atomic form and cause hydrogen embrittlement. This chapter outlines the factors which cause hydrogen embrittlement, its subsequent effects and failure mechanisms, and then elaborates on methods for reducing or eliminating the problem. Since steels are particularly prone to hydrogen embrittlement, emphasis is placed on these alloys. A section on permeation of hydrogen through various protective coatings is included to show the effectiveness of various barrier layers on minimizing hydrogen egress to substrates. Some excellent materials science investigative work showing how electroless copper deposits are embrittled by hydrogen is also presented along with data on hydrogen pick-up as a result of chemical milling of various steel and titanium alloys.

Hydrogen embrittlement is a generic term used to describe a wide variety of fracture phenomena having a common relationship to the presence of hydrogen in the alloy as a solute element or in the atmosphere as a gas (1). Louthan (2) lists the following problems as a result of hydrogen embrittlement and/or hydriding: failures of fuel cladding in nuclear reactors, breakage of aircraft components, leakage from gas filled pressure vessels used by NASA, delayed failure in numerous high strength steels, reductions in mechanical properties of nuclear materials, and blisters or fisheyes in copper, aluminum and steel parts. Steels, particularly those with high strengths of 1240 to 2140 MPa (180,000 to 310,000 psi) are prone to hydrogen embrittlement regardless of temperature (3)(4), (Figure 1). However, hydrogen embrittlement is not specific to just high strength steels.

11

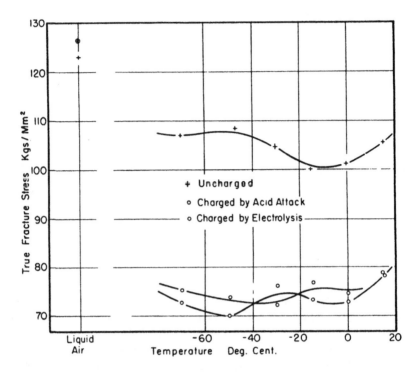

Figure 1: Fracture stress as a function of hydrogen absorption and temperature for 0.08% carbon steel. From reference 4. Reprinted with permission of ASM.

Nickel, titanium, aluminum (4)(5) and even electroless copper deposits (6) exhibit the phenomenon. It appears that any material can become embrittled by a pressure effect if hydrogen bubbles are introduced by a means such as electrodeposition and this state remains unchanged until hydrogen atoms escape from the bubbles (6). In some cases, the failure can be so abrupt and forceful as to seem almost explosive (7). Geduld reminisced that "one of the most spectacular and memorable sounds associated with zinc plating was standing in a quiet storage room next to drums of recently zinc plated steel springs and listening to the metallic shriek of self-destruction. as the springs slowly destroyed themselves in order to release occluded hydrogen" (8).

Some examples of the effects of hydrogen on the structure of metals are shown in Figures 2 through 10. Figure 2 is a photomicrograph of commercial copper that had been heated in hydrogen. The treatment produced a porous, degenerated structure of low strength and ductility. Figure 3 shows the wall of a heavy pressure vessel used in the petrochemical industry that developed large internal blisters and cracks from the action of hydrogen as a result of sulfide corrosion. Figure 4 shows a vanadium wire that literally shattered when it was cathodically charged with hydrogen in an electrolytic cell (9).

Figures 5 through 10 show the influence of hydrogen on steel (10). Figures 5 and 7 are a bar that was not cathodically treated, therefore, was not hydrogen embrittled. The same bar is shown in Figures 6 and 8 after cathodic treatment to introduce hydrogen and the severe damage is clearly evident. Figure 9 is a cathodically treated specimen showing partial relief as a result of heating at 200°C for 5 minutes while Figure 10 shows a sample exhibiting full relief as a result of heating at 200°C for 1 hour.

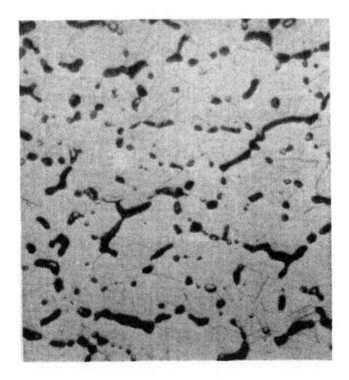

Figure 2: Structure of commercial copper after heating in hydrogen for 3 hours at 750°C (about 250 X).   From reference 9.   Reprinted with permission of Science.

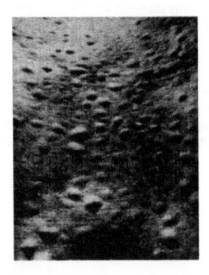

Figure 3: Hydrogen blisters in the wall of a steel container. From reference 9. Reprinted with permission of Science.

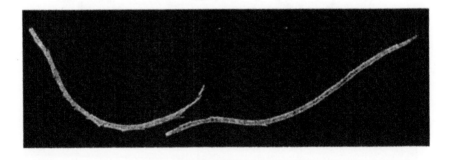

Figure 4: Vanadium wire shattered by cathodic charging with hydrogen. From reference 9. Reprinted with permission of Science.

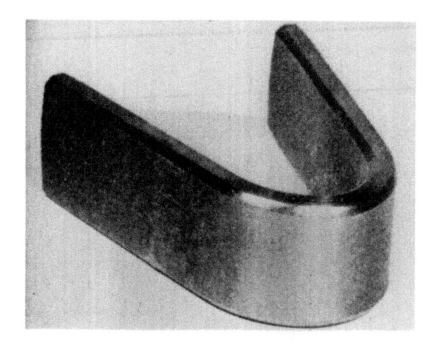

Figure 5: Steel sample with no cathodic treatment (× 1). From reference 10.

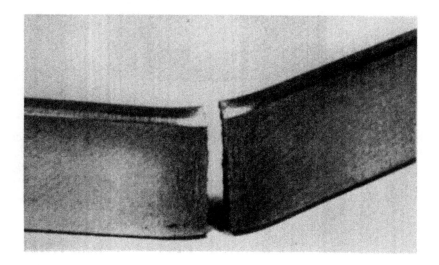

Figure 6: Steel sample after cathodic treatment (× 1). From reference 10.

Figure 7: Steel sample with no cathodic treatment (× 3). From reference 10.

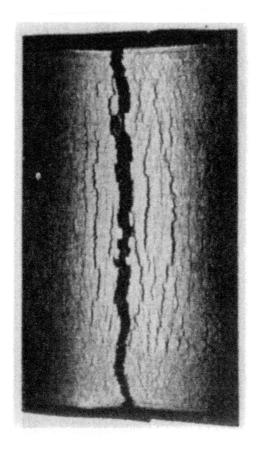

Figure 8: Steel sample after cathodic treatment (× 3). From reference 10.

Figure 9: Steel sample after cathodic treatment than heating at 200°C for five minutes (× 3). From reference 10.

Figure 10: Steel sample after cathodic treatment then heating at 200°C for one hour (× 3). From reference 10.

Platers must be especially aware of hydrogen embrittlement effects since many preplating and plating operations can be potent sources of absorbable hydrogen. Table 1, which lists sources of hydrogen, clearly shows that cathodic cleaning, pickling and electroplating are all culprits. This table also shows why anodic cleaning, which generates no hydrogen, is preferable to cathodic cleaning which generates copious amounts of hydrogen. Since corrosion reactions are also generators of hydrogen, care in choosing the proper coating to prevent corrosion is also quite important. In general, any process producing atomic hydrogen at a metal surface will induce considerable hydrogen absorption in that metal. However, not all the hydrogen atoms released at the surface enter metals; a large fraction combines or recombines to form bubbles of gaseous or molecular hydrogen which is not soluble in metals (9).

MECHANISM

A variety of mechanisms have been proposed to explain hydrogen embrittlement. In fact within a given system, depending on the source of hydrogen and the nature of the applied stress, the mechanism may change. The following are suggested by Birnbaum (1): "Non-hydride forming systems such as iron and nickel alloys which do not form hydrides under the conditions in which they are embrittled fail because hydrogen decreases the atomic bonding (decohesion). In many of these systems the fracture seems to be associated with hydrogen-induced plasticity in the vicinity of the crack tip. Metals such as niobium, zirconium, or titanium, which can form stable hydrides, appear to fracture by a stress-induced hydride formation and cleavage mechanism. Other mechanisms, such as adsorption-decreased surface energy and high-pressure hydrogen gas bubble formation, have also been suggested and may play a role in specific systems" (1).

STEELS

Hydrogen embrittlement effects are most pronounced in steels. These effects can take the form of reduced ductility, ease of crack initiation and/or propagation, the development of hydrogen-induced damage, such as surface blisters and cracks or internal voids, and in certain cases changes in the yield behavior (11). With steels, the problem occurs because of one or more of four primary factors: temperature, microstructure, tensile stresses, and hydrogen content (12). First, room temperature is just about right. An important consequence of the ease of interstitial diffusion (which is the way hydrogen moves about lattices) is the fact that considerable diffusion can

Table 1: Sources of Hydrogen

- ACID TYPE CORROSION

    $Fe + 2HCl \rightarrow FeCl_2 + H_2$

- ELECTROCHEMICAL CLEANING

    ANODE REACTION

    $4\ (OH)^- - 4e \rightarrow 2H_2O + O_2$

    CATHODE REACTION

    $4H^+ + 4e \rightarrow 2H_2$

- PICKLING

    $ACID + METAL \rightarrow SALT + H_2$

- ELECTROPLATING

    CATHODE REACTIONS

    $Ni^{+2} + 2e^- \rightarrow Ni\ METAL$

    $2H^+ + 2e^- \rightarrow H_2$

- CONTAINMENT VESSELS FOR H_2

- DAMPNESS IN MOLDS DURING CASTING

- HUMIDITY IN FURNACES DURING HEAT TREATING

- REMNANTS OF DRAWING LUBRICANTS

occur rather quickly even at low temperatures, in fact, even at room temperature. This is one of the circumstances which can lead to hydrogen pickup by metals during aqueous metal finishing operations at the comparatively low temperatures which comprise the range in which water is a liquid (13). Secondly, the microstructure of the plated part must be susceptible to the cracking mechanism and the martensite, bainite, and fine pearlite of quenched and tempered and cold drawn steels can allow the cracking mechanisms to operate at hardnesses down to Rockwell C 24 and lower. Third, tensile stresses are required and even if the part is loaded in compression, tensile stresses can develop and cracks will grow perpendicular to these local tensile stresses. Fourth, a certain amount of hydrogen is required (12).

PREVENTION OF HYDROGEN EMBRITTLEMENT

The prevention of hydrogen embrittlement failure requires a multi-pronged approach:

1. Prevention of hydrogen absorption wherever possible.

2. Elimination of the residual stress in the part before processing.

3. Baking after processing to remove absorbed hydrogen before it can damage the part.

Although beyond the control of the plater, hydrogen absorption can occur because of dampness in molds during casting or from humidity in furnaces during manufacture of the alloy. Baking parts before plating can help minimize stresses and remove absorbed hydrogen.   Table 2, a recommendation under development by ASTM, suggests times and tempera-

Table 2: Stress Relief Requirements for High Strength Steels (a)

Tensile Strength		Temp.	Time
MPa	psi	°C	Hours
1000-1400	145,000-203,000	200-230	Minimum 3
1401-1800	203,000-261,000	200-230	Minimum 18
Over 1800	Over 267,000	200-230	Minimum 24

a. From reference 14.

tures (14). Electropolishing before plating could also help; hydrogen entry into sensitive steels may be less than when the surface is stressed (15). Shot peening before plating has also been shown to reduce or even prevent the absorption of hydrogen (7). Other practical steps to minimize hydrogen embrittlement include (16)(17):

- Avoidance of cathodic cleaning, pickling or activation treatments whenever possible, by use of alkaline soak cleaning and anodic cleaning.

- Use of vapor degreasing or solvent cleaning to remove the bulk of grease, oil, or other contaminants before cleaning in aqueous solutions.

- Use of mechanical means (such as tumbling, sand , or grit blasting, vapor blasting, etc.) for oxide and scale removal, rather than pickling.

- Use of inhibited acid pickling solutions. The inhibitors either cut down on the amount of metal dissolved and thereby reduce the amount of hydrogen generated or they can change conditions at the surface so that less of the generated hydrogen enters the metal.

- Use of low embrittling electroplating processes such as special solution compositions and operating conditions which result in either a lower pickup of hydrogen or in a deposit that allows easier removal of the absorbed hydrogen during the baking treatment. Examples include 1) use of fluoborate or Cd-Ti instead of cyanide cadmium, 2) if using cyanide cadmium, plating to a thickness of 5 μm, baking for 3 hours at 190°C to remove hydrogen, and then continuing plating to the required thickness, or 3) use of a more permeable metal such as nickel where possible since hydrogen escapes through nickel far more easily than through zinc or cadmium.

- Use of coating techniques that avoid or minimize hydrogen embrittlement; e.g., vacuum deposited coatings, mechanical plating, and organic coatings.

- Baking after coating to provide hydrogen embrittlement relief. This is a standard practice for removing hydrogen from plated parts and verified in Figure 11 which shows the effect of baking at 149°C on the time-to-failure of notched specimens of 4340 steel heat treated to a strength level of 1590 MPa (230,000 psi). Table 3 lists baking recommendations suggested by ASTM (14).

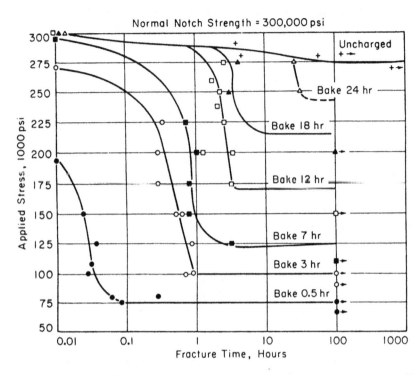

Figure 11: The effect of baking at 149°C on the time-to-failure at a given level of applied stress and stress ratio of notched specimens of 4340 steel heat treated to a strength level of 1586 MPa (230,000 psi). From reference 18. Reprinted with permission of ASM

Table 3: Baking After Plating Recommendations of ASTM (a)

Tensile Strength		Temp.	Time
MPa	psi	°C	Hours
1000-1100	145,000-160,000	190-220	Minimum 8
1101-1200	160,000-174,000	190-220	Minimum 10
1201-1300	174,000-189,000	190-220	Minimum 12
1301-1400	189,000-203,000	190-220	Minimum 14
1401-1500	203,000-218,000	190-220	Minimum 16
1501-1600	218,000-232,000	190-220	Minimum 18
1601-1700	232,000-247,000	190-220	Minimum 20
1701-1800	247,000-261,000	190-220	Minimum 22

a. From reference 14.

Statistically designed screening experiments were conducted to determine the significance of various parameters on the hydrogen content of bright and dull cadmium plated 4340 steel after baking (19)(20). Five variables were investigated: two plating batches, a delay between plating and baking, a delay between baking and measuring, the humidity conditions during this latter delay, and the baking time. Table 4 summarizes the data and shows that for bright cadmium, only the baking time was significant whereas with dull cadmium, the batch was also significant. The reason for this is probably that small changes in concentrations of the solution ingredients, especially carbonate, can have a significant effect on the nature of the deposit from a dull cadmium solution. A delay in baking of 24 hours had no effect on the final hydrogen content. Also, the hydrogen concentration was not altered if the specimens were held for a month at relative humidities of up to 50% after baking.

Table 4:   Plackett-Burman Results on Removal of Hydrogen From Bright and Dull Cadmium Plated 4340 Steel (a)

Column	Variable	Limits Bright − +		Factor effect	Limits Dull − +		Factor effect
x_1	Plating batch	2	1	0.087	1	2	0.145
x_2	Time between plating and baking (h)	1	24	0.127	1	24	0.038
x_3	Baking time (h)	8	48	0.350	8	20	0.172
x_4	Time between baking and measuring (h)	2	690	0.040	2	690	0.065
x_5	Relative humidity (%)**	0	35	0.077	0	50	0.005
	Minimum significant factor effect (90% confidence level)			0.230			0.093

**0% = desiccator; 35% = ambient; 50% = over saturated $Na_2 Cr_2 O_7$.

a. From reference 19.

Since many practitioners believe that a delay between plating and baking could be important, another experiment was run with just two variables, baking time and delay before baking. Bright cadmium plated specimens were baked for 3 and 72 hours, with delays before baking of 1/4 and 24 hours (20). Data in Table 5 show diffusable hydrogen concentration as a function of baking time and delay before baking. Results clearly reveal that there was no effect on the hydrogen concentration whether or not the baking was done as soon as possible after plating. In spite of these results it is possible that elapsed time between plating and baking can be sufficiently long enough that the migrating hydrogen reaches the critical concentration for crack initiation. No amount of baking will ever repair these cracks; the substrate will have a permanent reduction in yield strength (21).

Table 5: Two Variable, Two-Level Experimental Design and Results for Bright Cadmium Plated 4340 Steel (a)

Trial	Baking Time, h	Time, h	Hydrogen Concentration $\mu A/cm^2$	Avg.
1	3	1/4	0.88	1.07
			1.26	
2	3	24	1.08	1.05
			1.02	
3	72	1/4	0.26	0.28
			0.31	
4	72	24	0.28	0.30
			0.31	

a. From reference 20. Background level was 0.22 $\mu A/cm^2$

Cd-Ti Plating

Cd-Ti plating, an approach to inhibit hydrogen embrittlement, was introduced in the 1960's (22). This technique utilizes a standard cadmium cyanide solution with a sparsely soluble titanium compound plus hydrogen peroxide. When properly operated the deposit contains from 0.1 to 0.5% Ti. This process has been used for coating high strength landing gear actuation cylinders, linkage shafts and threaded rods subjected to high stress (23). A noncyanide electrolyte prepared by adding a predissolved Ti compound to a neutral ammoniacal cadmium solution is also available (24). With this electrolyte, fine-grained Cd-Ti deposits containing 0.1 to 0.7% Ti have been obtained. It is reported that with respect to throwing power, corrosion protection and hydrogen embrittlement, the noncyanide solution is better than the cyanide solution. The Ti compound is stable in the noncyanide solution, so the continuous filtration and frequent analysis required with the Cd-Ti cyanide process are avoided. The process has been used since 1975 for applying protective coatings on high strength structural steel, spring wire and high quality instrument steel (24). Figure 12, which shows hydrogen permeation data for a noncyanide Cd-Ti solution, clearly reveals the influence of Ti in inhibiting hydrogen absorption.

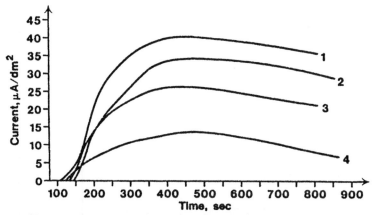

Figure 12: Hydrogen penetration current vs. time in Cd plating solution with (1) no Ti, (2) 0.067 g/l Ti, (3) 2.2 g/l Ti, and (4) 3.1 g/l Ti. From reference 24. Adapted from reference 24.

Mechanical Plating

Mechanical plating is one of the coating techniques available for minimizing hydrogen embrittlement. Also known as peen plating, mechanical plating is an impact process used to apply deposits of zinc, cadmium or tin. It has been a viable alternative to electroplating for the application of sacrificial metal coatings on small parts such as nails, screws, bolts, nuts, washers and stampings for over 30 years (25). Table 6 includes static test data for 1075 steel heat treated to Rc 52-55 before being electroplated with 12.5 μm (0.5 mil) cadmium by normal procedures or by mechanical plating. Parts coated by mechanical plating exhibited no hydrogen embrittlement, whereas, those coated by normal plating exhibited

Table 6: Stress-Endurance Static Tests for Cadmium Plated 1075 Steel (Rc 52-55) Parts (a)

Type of Plating	Group	Less Than 5 Min.	Between 5 min. & 1 hr.	Over one to 240 hrs.	Total No. of Pieces Failed
Electro-plated	Small rings austempered	9	12	One (between 1 & 2 hrs.)	30
	Quench and Tempered	30	--	--	30
	Large rings austempered	0	0	0	0
	Quench and tempered	4	23	One(Between 24 & 120 hrs.)	28
Mechanically plated	All groups	0	0	0	0

a.   Thirty parts were tested for each condition. For all tests, plating thickness was 12.5 μm (0.5 mil). These data are from reference 26.

various degrees of failure, ranging from 100% failure for small rings which had been quenched and tempered to no failure for large rings which had been austempered. Dynamic testing did reveal that some embrittlement occurred as a result of the mechanical plating process although not as extensive as that obtained with normal plating (26).

Physical Vapor Deposition

One coating technique that eliminates the potential of hydrogen embrittlement is that of physical vapor deposition (PVD), particularly ion plating. PVD processes such as evaporation, sputtering and ion plating are discussed in some detail in the chapter on Adhesion. Since these processes are done in vacuum, the chance of embrittlement by hydrogen is precluded. For production parts, precleaning consists of solvent cleaning followed by mechanical cleaning with dry aluminum oxide grit (27). Therefore, there is no need for costly embrittlement relief procedures nor is there the risk of catastrophic failure due to processing. Ion plated aluminum coatings have been used for over 20 years particularly for aircraft industry applications (28). This aluminum deposit protects better than either electroplated or vacuum deposited cadmium in acetic salt fog and most outdoor environments. Class I coatings, 25 µm (0.001 inch minimum) of ion vapor deposited aluminum have averaged 7500 hours before the formation of red rust in 5 percent neutral salt fog when tested under ASTM-E-117 (29).

Permeation

Since one of the key methods for minimizing hydrogen embrittlement is the use of a barrier coating, the influence of various coatings on the permeability of hydrogen is of importance. Thin layers of either Pt, Cu, or electroless nickel decrease permeability of hydrogen through iron (30). The coatings do not have to be thick or even continuous to be effective suggesting that a catalytic mechanism is responsible for the marked reduction in hydrogen permeation through the iron. Au (31)(32), Sn and Sn-Pb alloy coatings are also very effective permeation barriers (33)-(35). Lead coatings are effective in preventing hydrogen cracking on a variety of steels in many different environments (36)-(38). Permeation data presented in Figures 13 through 15 show that:

- A Pt coating of only 0.015 µm was very effective in reducing hydrogen permeation through iron (Figure 13).

- Cu was noticeably more effective than Ni in reducing the rate of hydrogen uptake by iron (Figure 14).

-    With 1017 steel, brush plating with 70Pb-30Sn noticeably
     reduced the permeability (Figure 15). An imperfect brush
     plated zinc coating was also quite effective in reducing
     permeability.

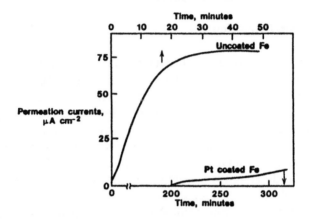

Figure 13: Effect of a platinum coating (0.015 μm thick) on the permeation
of hydrogen through Ferrovac E iron membranes. Charging current density
was 2 mA/cm². Charging solution was 0.1 N NaOH plus 20 ppm As^{3+}.
Adapted from reference 30.

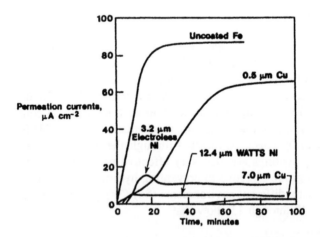

Figure 14: Effect of copper, nickel and electroless nickel coatings on the
permeation of hydrogen through Ferrovac E iron membranes. Charging
current density was 2 mA/cm². Charging solution was 0.1 N NaOH plus 20
ppm As [3+]. Adapted from reference 30.

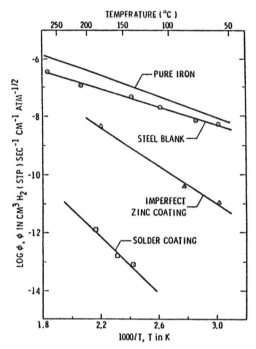

Figure 15: Brush plating as a means of reducing hdyrogen uptake and permeation in 1017 steel. Adapted from reference 33.

Extensive work for NASA has shown the effectiveness of Cu and Au in reducing the permeability of hydrogen. For example, electrodeposited nickel is highly susceptible to hydrogen environment embrittlement (HEE) (32)(39)(40). Both ductility and tensile strength of notched specimens show reductions up to 70 percent in 48.3 Mpa (7000 psi) hydrogen when compared with an inert environment at room temperature. Annealing at 343°C minimizes the HEE of electrodeposited nickel regardless of the current density used to deposit the nickel. Another approach to prevent HEE of electrodeposited nickel is to coat the nickel with copper or gold. Tensile tests conducted to determine the effectiveness of 80 μm thick copper and 25 μm thick gold are summarized in Table 7. Both coatings allowed the electrodeposited nickel to retain its ductility in high pressure hydrogen (32).

Since metallurgically prepared nickel alloys are also notoriously susceptible to hydrogen embrittlement, NASA utilizes an electrodeposited copper layer (150 um) to protect the inner surface of a four ply nickel alloy bellows from contacting a hydrogen atmosphere. This bellows is used in the Space Shuttle engine turbine drive and discharger ducts prior to forming (41).

Table 7: Effect of Copper and Gold Coatings on the Tensile Properties of Unnotched Electrodeposited Nickel Specimens in High-Pressure Hydrogen At Room Temperature (a)

Coating	Test Environ.	Test Pressure		Ultimate Tensile Strength		Elongation (%)	Reduction Area (%)
		MPa	psi	MPa	psi		
Uncoated	He	8.3	1,200	538	78,000	23	90
Uncoated	H$_2$	8.3	1,200	503	73,000	5	30
Copper(b)	H$_2$	48.3	7,000	572	83,000	25	91
Gold (c)	H$_2$	8.3	1,200	531	77,000	25	89

a. These data are from reference 32.
b. Copper was 75 μm (3 mil) thick.
c. Gold was 25 μm (1 mil) thick.

Electroless Copper

An excellent application of materials science principles is the work by researchers at AT & T Bell Laboratories on electroless copper. By utilizing a variety of sophisticated analytical techniques including inert gas fusion analysis, ion microprobe analysis, thin film ductility measurements, and scanning and transmission electron microscopy they showed that hydrogen is responsible for the lower ductility noted in electroless copper deposits when compared with electrodeposited copper films (6)(42)-(49). They attributed this ductility loss to hydrogen embrittlement contrary to the common notion that physical properties of Group IB metals (copper, silver, and gold) are insensitive to hydrogen (44). This work should be generally applicable to other electrodeposited and electroless films in which the deposition process involves a simultaneous discharge of both metal and hydrogen ions (6).

Electroless copper deposition is used extensively in the fabrication of printed wiring boards. Since these deposits are often subjected to a hot solder bath during the printed wiring board manufacturing process, good ductility is required to withstand thermal shock. An item of concern with electroless copper deposits is their ductility which is generally much poorer ($\sim$ 3.5%) than that of electrolytic copper (12.6 to 16.5%) (6). This loss in film ductility for electroless copper deposits has been attributed to a high (10^4 atm.) pressure developed because of hydrogen gas bubbles in analogy to the pressure effect in classical hydrogen embrittlement (6). In the electroless copper deposition process, the formation of hydrogen gas is an integral part of the overall deposition reaction:

$$Cu(II) + 2HCHO + 4OH \rightarrow Cu + 2HCOO + 2H_2O + H_2$$

Some of the hydrogen atoms and/or molecules can be entrapped in the deposit in the form of interstitial atoms or gas bubbles (48). By contrast, in the case of electrolytic copper deposition, hydrogen evolution can be avoided by choosing the deposition potential below the hydrogen overpotential to prevent hydrogen reduction. This cannot be done with electroless copper deposition since hydrogen reduction is an integral part of the deposition reaction.

Table 8, which lists the concentration ranges of impurity elements found in an electroless copper deposit, shows that hydrogen content is disproportionately high compared to the other elements (46). Some of this hydrogen can be removed by annealing at relatively low temperatures and this results in an improvement in ductility. Figure 16 shows the variation of ductility and hydrogen content with annealing time at 150°C in nitrogen. The ductility improves with annealing time and reaches a nearly constant

Table 8: Inclusions in Electroless Copper Deposits (a)

Element	ppm, Weight	ppm, Atomic
H	30-200	1900-12700
C	90-800	480-4230
O	70-250	280-990
N	20-110	90-500
Na	20-70	55-190

a. These data are from reference 46.

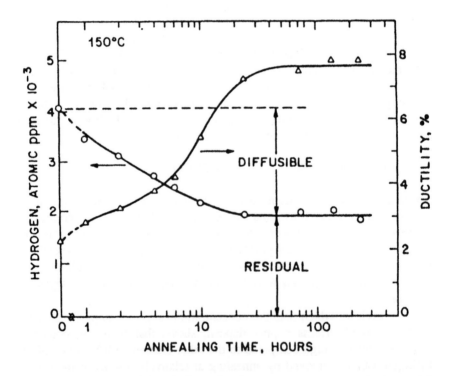

Figure 16: Variation of hydrogen content and ductility with annealing time at 150°C for an electroless copper deposit. From reference 46. Reprinted with permission of The Electrochemical Soc.

level after 24 hours. In somewhat similar fashion, the hydrogen content decreases initially and becomes constant after the same length of time. Inspection of the hydrogen curve reveals that two kinds of hydrogen are present in the deposit, "diffusable hydrogen" which escapes on annealing, and "residual hydrogen" which is not removed by annealing (46). The close correlation between the loss of hydrogen and improvement in ductility shown in Figure 16 is further demonstrated by a cathodic charging experiment in which an annealed deposit containing no diffusable hydrogen was made a cathode in an acidic solution to evolve hydrogen for an extended period of time, and the diffusable hydrogen and ductility remeasured (46). The results are presented in Table 9.

This extensive work by researchers at AT & T Bell Laboratories has led them to conclude that hydrogen inclusion introduces two sources of embrittlement into electroless copper. The first one is the classical hydrogen embrittlement by the pressure effect and the second is the introduction of void regions, which promote the ductile fracture by the void coalescence mechanism. The former embrittlement can be removed by annealing at 150°C but the latter remains constant (47).

Table 9: Cathodic Charging Experiment With Electroless Copper Deposits (a)

Condition	Ductility %	Diffusable Hydrogen ppm, Atomic
As deposited	2.1	2780
After annealing (b)	6.5	0
After charging	3.8	2360
After reannealing (b)	6.4	0

a.   These data are from reference 46. Cathodic charging conditions: 0.05M H_2SO_4, 0.001M As_2O_3, (10mA/cm², 15 hours)

b.   Annealing was done at 150°C for 24 hours.

Chemical Milling

Chemical milling has evolved as a valuable complement to conventional methods of metal removal. Any metal that can be dissolved chemically in solution can be chemically milled. Aluminum, beryllium, magnesium, titanium, and various steel and stainless steel alloys are among those most commonly milled although refractory metals such as molybdenum, tungsten, columbium, and zirconium, can also be handled. Parts can be flat, preformed, or irregular, and metal can be removed from selected areas or the entire surface (50).

Chemical milling of steels, stainlesses and high-temperature alloys typically requires extremely corrosive raw-acid mixtures. In spite of the fact that much hydrogen is generated during the process, milled parts suffer little or no degradation. Data in Table 10 summarize the influence of chemical milling on the tensile properties of various alloys. In most cases, no degradation was noted. With 4340, some embrittlement was obtained with chemically milled specimens, but properties were restored by aging at room temperature.

Titanium alloys are chemically milled primarily to provide a maximum strength-to-weight ratio. As with steels, various acids are used for milling,therefore, hydrogen is generated (50)(59)-(61). Since titanium and its alloys are susceptible to hydrogen embrittlement (Figure 17) the amount of hydrogen picked up when these materials are chemically milled is of major concern (2). In titanium structures, hydrogen can concentrate at the surface causing a reduction in surface sensitive properties. The most important factors governing the amount of hydrogen absorbed are the composition and metallurgical structure of the alloy, the composition and temperature of the etching solution, the etching time, the sequence in which the parts fit into the milling cycle, whether the parts are etched on one or both sides, and the mass of material remaining after etching. For example, hydrogen pickup is much greater when specimens are milled from two sides rather than just one. Figure 18 contains data for Ti-6Al-6V-2Sn, showing that absorption is a function of the ratio of chemically milled surface to final volume and not of the amount of metal removed by milling (59). Table 11 summarizes data on hydrogen absorption for various titanium alloys.

TESTS FOR HYDROGEN EMBRITTLEMENT

A variety of tests are available for assessing hydrogen embrittlement but these will not be covered here. For those interested in these tests, references 63 and 64 provide a good starting point.

Table 10: Influence of Chemical Milling on the Tensile Properties of Some Alloy Steels

Alloy	Tests	Results	Ref.
4340, D6ac 4349, H-11, 17-7 PH	Sustained Load (1800 MN/m²) Sustained Load	No Hydrogen embrittlement. Embrittled, but recovery occurred with 1 week after storage at room temperature or after 49 hours at room temperature followed by a 4 hour bake at 190°C.	51
4340	Bend	Removal of 250 or 500 μm by chemical milling resulted in hydrogen embrittlement. Ductility was restored by aging at room temperature for 8 to 33 hours.	52
1Cr-Mo Steel 3CR-Mo-V Steel 5CR-Mo-V Steel	Sustained Load	No adverse effects.	53
18% Ni Maraging Steel	Sustained Load (2000 MN/m²)	No adverse effects.	54
301 Stainless	Tensile	No adverse effects.	55
Rene 41	Tensile	Higher strengths than for base-metal controls.	51
Rene 41	Tensile	Slight reduction in tensile strength at 760°C, no adverse effects between 760°C and 980°C. Some reduction in elongtion: 15% at 27°C, 40% at 760°C and 20% at 980°C.	56
Inconel X	Tensile and Stress—Rupture	No adverse effects.	57
PH 14-8	Tensile	Some reduction in properties, but this was attributed to differences in specimen thickness.	58

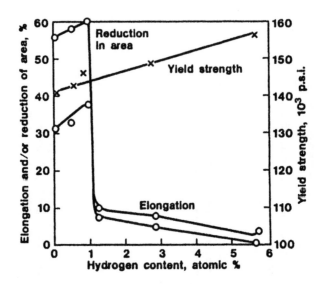

Figure 17: Tensile properties of Ti-8Mn as a function of hydrogen content. Adapted from reference 4.

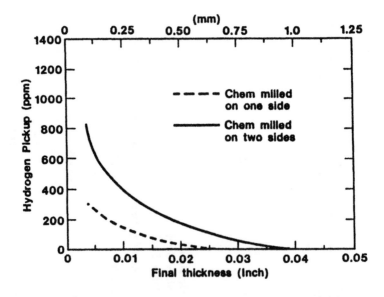

Figure 18: Hydrogen absorption vs. depth of cut for chemically milled Ti-6Al-6V-2Sn. Adapted from reference 59.

Hydrogen Embrittlement 39

Table 11: Influence of Chemcial Milling on the Hydrogen Content of Various Titanium Alloys

Alloy	Metal Removed		Hydrogen, ppm			Ref.
	μm	inch	Before	After	Change	
Ti-6Al-4V	125	.005	50	46	-4	59
	250	.010	29	39	10	
	500	.020	29	48	19	
	750	.030	29	68	29	
	750	.030	78	36	-42	
Ti-5Al-2.5Sn	250	.010	34	37	3	59
	500	.020	34	35	1	
	750	.030	34	36	2	
Ti-8Al-1Mo-1V	250	.010	46	39	-7	59
	500	.020	46	45	-1	
	750	.030	46	42	-4	
Ti-15V-3Cr-3Al-2Sn	5	.0002	-	-	0	62
	35	.0014	-	-	11	
	75	.0030	-	-	50	

REFERENCES

1.   H.K. Birnbaum, "Hydrogen Embrittlement", *Encyclopedia of Materials Science and Engineering*, M.B. Bever, Editor, Pergamon Press, 2240 (1986)

2.   M.R. Louthan, Jr., "The Effect of Hydrogen on Metals", *Corrosion Mechanisms*, F. Mansfeld, Editor, Marcel Dekker Inc., NY (1987)

3.   E.A. Groshart, "Design and Finish Requirements of High Strength Steels", *Metal Finishing* 82, 49 (March 1984)

4.   P. Bastien and P. Azou, "Effect of Hydrogen on the Deformation and Fracture of Iron and Steel in Simple Tension", *Proceedings of the First World Metallurgical Congress*, ASM (1951)

5.   D. Nguyen, A.W. Thompson and I.M. Bernstein, "Microstructural Effects on Hydrogen Embrittlement in a High Purity 7075 Aluminum Alloy", *Acta. Metall.*, 35, 2417 (1987)

6.   S. Nakahara and Y. Okinaka, "Microstructure and Ductility of Electroless Copper Deposits", *Acta Metall.*, 31, 713 (1983)

7.   L.J. Durney, "Hydrogen Embrittlement: Baking Prevents Breaking", *Products Finishing*, 49, 90 (Sept 1985)

8.   H. Geduld, *Zinc Plating*, ASM International, 203 (1988)

9.   H.C. Rogers, "Hydrogen Embrittlement", *Science*, 159, 1057 (1968)

10.   L.E. Probert and J.J. Rollinson, "Hydrogen Embrittlement of High Tensile Steels During Chemical and Electrochemical Processing", *Electroplating and Metal Finishing*, 14, 396 (Nov 1961)

11.   I.M. Bernstein and A.W. Thompson, "Hydrogen Embrittlement of Steels", *Encyclopedia of Materials Science and Engineering*, M.B. Bever, Editor, Pergamon Press, 2241 (1988)

12.   C.D. Beachem, "Mechanisms of Cracking of Hydrogen-Charged (Hydrogen-Embrittled) Aerospace Materials", *Proceedings AESF Aerospace Symposium* (Jan 1989)

13.   H.J. Read, "The Metallurgical Aspects of Hydrogen Embrittlement in Metal Finishing", *47th Annual Technical Proceedings*, AES, 110 (1960)

14.   A.W. Grobin, "Other ASTM Committees and ISO Committees Involved in Hydrogen Embrittlement Test Methods", *Hydrogen Embrittlement: Prevention and Control*, ASTM STP 962, L. Raymond, Editor, ASTM, Philadelphia, PA, 46 (1988)

15.   C.L. Faust, "Electropolishing Carbon and Low Alloy Steels-Part 1", *Metal Finishing*, 81, 47 (May 1983)

16.   E.T. Clegg, "Hydrogen Embrittlement Coverage by U. S. Government Standardization Documents", *Hydrogen Embrittlement: Prevention and Control*, ASTM STP 962, L. Raymond, Editor, ASTM, Philadelphia, PA, 37 (1988)

17.   A.W. Thompson, "Metallurgical Characteristics of Hydrogen Embrittlement", *Plating & Surface Finishing*, 65, 36 (Sept 1978)

18.   A.R. Troiano, "The Role of Hydrogen and Other Interstitials on the Mechanical Behavior of Metals", *Trans. American Society for Metals*, 52, 54 (1960)

19.   D.A. Berman, "Removal of Hydrogen From Cadmium Plated High Strength Steel by Baking-A Statistically Designed Study", Report No. NADC-82238-60 (Nov 1982), Naval Air Development Center, Warminster, PA

20.   D.A. Berman, "The Effect of Baking and Stress on the Hydrogen Content of Cadmium Plated High Strength Steels", AD-A166869 (Dec 1985), Naval Air Development Center, Warminster, PA

21.   A.W. Grobin, Jr., "Hydrogen Embrittlement Problems", *ASTM Standardization News*, 18, 30 (March 1990)

22.   K. Takata, Japanese patents SHO-35 18260 (1960) and SHO-38 20703 (1963)

23.   AMS-2419A, "Cadmium-Titanium Alloy Plating", *Soc. of Automotive Engineers*, Warrendale, PA (1979)

24.   W. Sheng-Shui, C. Jing-Kun, S. Yuing-Mo and L. Jin-Kuel, "Cd-Ti Electrodeposits From a Noncyanide Bath", *Plating & Surface Finishing*, 68, 62 (Dec 1981)

25.   R.N. Holford, Jr., "Five-Year Outdoor Exposure Corrosion Comparison", *Metal Finishing*, 86, 17 (July 1988)

26.   L. Coch, "Plating Fasteners, Avoiding Embrittlement", *Products Finishing*, 51, 56 (May 1987)

27.   V.L. Holmes, D.E. Muehlberger and J.J. Reilly, "The Substitution of IVD Aluminum for Cadmium", ESL-TR-88-75, *Engineering & Services Laboratory*, Tyndall Air Force Base, FL (Aug 1989)

28.   E.R. Fannin and D.E. Muehlberger, "Ivadizer Applied Aluminum Coating Improves Corrosion Performance of Aircraft", McDonnell Aircraft Company, MCAIR 78-006 (1978)

29.   D.E. Muehlberger, "Ion Vapor Deposition of Aluminum-More Than Just a Cadmium Substitute", *Plating & Surface Finishing*, 70, 24 (Nov 1983)

30.   S.S. Chatterjee, B.G. Ateya and H.W. Pickering, "Effect of Electrodeposited Metals on the Permeation of Hydrogen Through Iron Membranes", *Metallurgical Transactions A*, 9A, 389 (1978)

31.   T-P. Perng, M.J. Johnson and C.L Alstetter, "Hydrogen Permeation Through Coated and Uncoated Waspalloy", *Metallurgical Transactions A*, 19A, 1187 (1988)

32.   W.T. Chandler, R.J. Walter, C.E. Moeller and H.W. Carpenter, "Effect of High-Pressure Hydrogen on Electrodeposited Nickel", *Plating & Surface Finishing* 65, 63 (May 1978)

33.   S.L. Robinson, W.A. Swansiger and A.D. Andrade, "The Role of Brush Plating in Future Hydrogen Storage and Transmission Systems", *Plating & Surface Finishing* 66, 46 (August 1979)

34.   D.R. Begeal, "The Permeation and Diffusion of Hydrogen and Deuterium Through Rodar, Tin-Coated Rodar and Solder-Coated Rodar", *J. Vac. Sci. Technol.*, 12, 405 (Jan/Feb 1975)

35.   J. Bowker and G.R. Piercy, "The Effect of a Tin Barrier on the Permeability of Hydrogen Through Mild Steel and Ferritic Stainless Steel", *Metallurgical Transactions A*, 15A, 2093 (1984)

36.   L. Freiman and V. Titov, "The Inhibition of Diffusion of Hydrogen Through Iron and Steel by Surface Films of Some Metals", *Zhur. Fiz. Khim.*, 30, 882 (1956)

37.   I. Matshushima and H.H. Uhlig, "Protection of Steel From Hydrogen Cracking by Thin Metallic Coatings", *J. Electrochem. Soc.*, 113, 555 (1966)

38.   H. Tardif and H. Marquis, "Protection of Steel From Hydrogen by Surface Coatings", *Canadian Metallurgical Quarterly*, 1, 153 (1962)

39.   R.P. Jewett, R.J. Walter, W.T. Chandler and R.P. Frohmberg, "Hydrogen Environment Embrittlement of Metals", *NASA CR-2163*, (Mar 1963)

40.   W.T. Chandler and R.J. Walter, "Hydrogen Embrittlement", *ASTM STP 543*, American Society for Testing and Materials, Philadelphia, PA, 182 (1974)

41.   "Electroplating Offers Embrittlement Protection", *NASA Tech Briefs*, p 140 (Spring 1979)

42.   S. Nakahara and Y. Okinaka, "Transmission Electron Microscopic Studies of Impurities and Gas Bubbles Incorporated in Plated Metal Films", *Properties of Electrodeposits: Their Significance and Measurement*, R. Sard, H. Leidheiser, Jr., and F. Ogburn, Editors, The Electrochemical Soc., Pennington, NJ, (1975)

43.   Y. Okinaka and S. Nakahara, "Hydrogen Embrittlement of Electroless Copper Deposits", *J. Electrochem. Soc.*, 123, 475 (1976)

44.   S. Nakahara and Y. Okinaka, "On the Effect of Hydrogen on Properties of Copper", *Scripta Metall.*, 19, 517 (1985)

45.   J.E. Graebner and Y. Okinaka, "Molecular Hydrogen in Electroless Copper Deposits", *J. Appl. Physics*, 60, 36 (July 1986)

46.   Y. Okinaka and H.K. Straschil, "The Effect of Inclusions on the Ductility of Electroless Copper Deposits", *J. Electrochem. Soc.*, 133, 2608 (1986)

47.   S. Nakahara, "Microscopic Mechanism of the Hydrogen Effect on the Ductility of Electroless Copper", *Acta Metall.*, 36, 1669 (1988)

48.   S. Nakahara and Y. Okinaka, "The Hydrogen Effects in Copper", *Materials Science and Engineering*, A, 101, 227 (1988)

49.   S. Nakahara., Y. Okinaka, and H.K. Straschil, "The Effect of Grain Size on Ductility and Impurity Content of Electroless Copper Deposits", *J. Electrochem. Soc.*, 136, 1120 (1989)

50.   J.W. Dini, "Fundamentals of Chemical Milling", *American Machinist*, 128, 113 (July 1984)

51.   C. Micillo, "Advanced Chemical Milling Processes", AFML-TR-68-237, or AD 847070 (August 1968)

52.   R.L. Jones, "A New Approach to Bend Testing for the Determination of Hydrogen Embrittlement of Sheet Materials", AD 681765 (June 1961)

53.   Anon., "The Chemical Contouring of 3% Chromium--Molybdenum-Vanadium and 5% Chromium-Molybdenum-Vanadium High Strength Steel", *Bristol Aerojet England*, BR-ARC-CP-811 (March 1964)

54.   E.C. Kedward and P.F. Langstone, "Chemical Contouring of 18 Percent Nickel Maraging Steel", *Sheet Metal Industries*, London, 46, 473 (June 1969)

55.   R.L. Jones and P. Bergstedt, "Compilation of Materials Research Data, Fourth Quarterly Progress Report-Phase 1", *General Dynamics*, AD 273065 (Dec 1961-Feb 1962)

56.   E. Howells, "Taper Chemical Milling of Rene 41 Tubes", Boeing Co., Seattle, Wash., MDR-2-14969 (March 1962)

57.   S.J. Ketcham, "Chemical Milling of Alloy Steels", Naval Air Engineering Center, NAEC-AML-2418 or AD 631952 (March 1966)

58.   B. Chapman and J. Derbyshire Jr., "Confirmation of the Close Tolerance Chem-Mill Process to PH 14-8 Mo Steel for the Apollo Heat Shield", North American Aviation, SDL 435 (Nov. 6, 1963)

59.   C. Micillo and C.J. Staebler Jr., "Chemical Milling Using Cut Maskant Etching Masks", *Photochemical Etching*, 3, 4 (June 1968)

60.   CHEMICAL MILLING, W.T. Harris, Oxford Univ. Press, New York (1976)

61.   J.W. Dini, "Chemical Milling", *International Metallurgical Reviews*, 20, 29 (1975)

62.   R. Messler Jr. and J. Masek, "Thermal Processing, Cleaning, Descaling and Chemical Milling of Ti-15V-3Cr-3Al-3Sn Titanium Alloy", TMS/AIME Paper F80-15 (1979)

63.   L. Raymond, "Evalaution of Hydrogen Embrittlement", *Metals Handbook, Ninth Edition*, Vol 13, Corrosion (1987), American Society for Metals, Metals Park, Ohio

64.   *Hydrogen Embrittlement: Prevention and Control*, ASTM STP 962, L. Raymond, Editor, American Society for Testing and Materials, Philadelphia, PA (1988)

3

ADHESION

INTRODUCTION

Adhesion refers to the bond (chemical or physical) between two adjacent materials, and is related to the force required to effect their complete separation. Cohesive forces are involved when the separation occurs within one of them rather than between the two (1). The ASTM defines adhesion as the "condition in which two surfaces are held together by either valence forces or by mechanical anchoring or by both together" (2). Adhesion is a macroscopic property which depends on three factors: 1) bonding across the interfacial region, 2) type of interfacial region (including amount and distribution of intrinsic stresses) and 3) the fracture mechanism which results in failure (3-5). Equating adhesion, which is a gross effect, to bonding or cleanliness may be very misleading. Failure of adhesion may be more related to fracture mechanisms than to bonding. In thin films, the intrinsic stress may result in adhesive failure even though chemical bonding may be high. Also, the interfacial morphology may lead to easy fracture though bonding is strong. With copper, if an acid dip prior to plating is too strong so that the etching results in the development of large areas with {111} planes constituting the surface, the subsequently deposited films may not only grow non-epitaxially, but also lose adhesion to the substrate forming an interfacial crack because of the voids (6). Good adhesion is promoted by: 1) strong bonding across the interfacial region, 2) low stress gradients, either from intrinsic or applied stress, 3) absence of easy fracture modes, and 4) no long term degradation modes (3-5).

Adhesion of a coating to its substrate is critical to its function. Mechanical, chemical, and metallurgical factors may contribute to such adhesion. For a coating to be retained and to perform its function, its adhesion to the substrate must tolerate mechanical stresses and elastoplastic distortions, thermal stress, and environment or process fluid displacement. Good adhesion performance of a coating depends on a variety of the attributes of the interface region, including its atomic bonding structure, its elastic moduli and state of stress, its thickness, purity and fracture toughness (7).

The durability of coatings is of prime importance in many applications and one of the main factors that govern this durability is adhesion. This is particularly true if the coating or substrate, is subject to corrosion or to a humid atmosphere, as under these circumstances any tendency for the film to peel from the substrate may well be aggravated. When adhesion is poor, rubbing action can cause localized rupture at the coating/substrate interface, leading to blistering or even complete spalling off of the coating. For example, material loss in wear tests was minimum with Pb/Sn films deposited by ion plating which results in very good adherence. By comparison, heavy material loss was obtained with Pb/Sn films deposited by evaporation which provides considerably less adherence. With the less adherent films deposited by evaporation, several failure mechanisms such as plucking, peeling, film displacement, etc., were observed (8).

In general, adhesion can be broken down into the following categories (9):

1.   *Interfacial adhesion*: the adhesive forces are centered around a narrow well defined interface, with minimal atomic mixing, such as gold on silica.

2.   *Interdiffusion adhesion*: the film and substrate diffuse into one another, over a wider interfacial region. For example, gold, evaporated onto freshly etched silicon (removing the surface oxide layer) at 50°C produces a diffuse interface extending many atomic layers.

3.   *Intermediate layer adhesion*: in many cases the film and substrate are separated by one or more layers of material of different chemical composition, as in the case of films deposited on unetched silicon whose surface is covered with several nanometers of oxide.

4.   *Mechanical interlocking*: this will occur to some degree wherever the substrate surface is not atomically flat and will account for some degree of random fluctuation of adhesive forces.

TESTING

Adhesion tests can be broken down into two categories, qualitative and quantitative. They vary from the simple scotch tape test to complicated flyer plate tests which require precision machined specimens and a very expensive testing facility. It is not the intent to provide a complete review of all adhesion tests in this chapter but rather provide some coverage of those that were used to generate the data that is presented later. For those interested in more detail, references 1 and 10-14 are recommended.

Table 1 gives a general breakdown of adhesion tests, classifying them into qualitative and quantitative. In many cases, the qualitative tests are quite adequate and are certainly easier and cheaper to perform. As with all tests, thickness of the coating can noticeably influence the results. This is shown in Figure 1 for the scotch tape test. Aluminum panels were not given any special activation treatment prior to plating with varying thicknesses of palladium so it was known that adhesion would be poor. The panel coated with only 1.25 µm (0.05 mil) of Pd indicates fairly good adhesion; only a small amount of coating was removed by the tape test. As the thickness of Pd was increased increasing amounts of coating were removed by the tape. Although not shown, if the coating were increased to around 25 µm (1 mil) no coating would be removed since the coating would be stronger than the tape even though the deposit would still be non-adherent. Likewise, with a very thin coating, e.g., around 0.5 µm (0.02 mil), no failure would be noticed with the scotch tape test. This strongly shows that with a qualitative test, a variety of results can be obtained and they can be quite misleading.

In cases where coatings are required for engineering applications, qualitative tests are often inadequate and must be replaced with tests that provide quantitative date. Of those listed in Table 1, four that were used to generate data that will subsequently be discussed include tensile, shear, peel and flyer plate so some details will be given for these tests.

A.   Conical Head Tensile Test

With this test, the electrodeposit, the substrate and the bond between the two are tested in a tensile fashion, the bond being normal to the loading direction. Flat panels are plated on both sides with thick electrodeposit (e.g., around 3 mm) and then conical head specimens are machined and tested using standard tension testing procedures. Figure 2 is a schematic of conical head tension specimens. More detail on this test can be found in references 15-17.

Table 1 - Adhesion Tests

Qualitative	Quantitative
Scotch tape	Tensile
Bending	Shear
Abrasion	Peel
Heating	Ultrasonics
Scribing	Centrifuge
Grinding	Flyer Plate
Impacting	

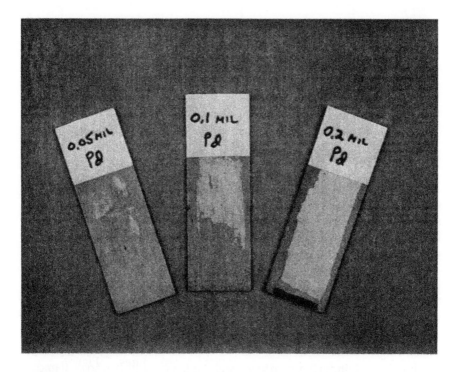

Figure 1: Scotch tape test for palladium plated on aluminum.

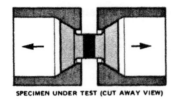

SPECIMEN UNDER TEST (CUT AWAY VIEW)

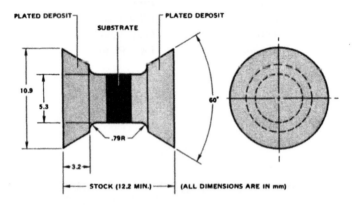

Figure 2: Conical head tensile specimen.

B.   Ring Shear Test

Ring shear tests are an effective, relatively simple method for obtaining quantitative data on the bond between coatings and their substrates. An added benefit with this type of test is that substrate material is easier to obtain and specimens cost less to fabricate and evaluate than for other types of quantitative tests (17).

A typical test is accomplished by preparing a cylindrical rod via the process under evaluation and then plating to a thickness of about 1.5 mm. The rod is machined in a manner that removes all of the plated deposit except for small rings of plating of predetermined width (generally 1.5 mm wide, spaced approximately 2.5 cm apart). The rod is then cut between the plated rings. These sections of the rod with the plated rings are tested by forcing the rod through a hardened steel die having a hole whose diameter is greater than that of the rod but less than that of the rod and the coating. The bond strength can be calculated by using the load to cause failure and the area of the coating. Figure 3 shows a ring shear test specimen and die. References 16-18 provide more detail on ring shear testing.

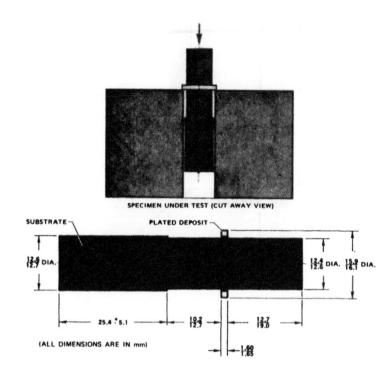

Figure 3: Ring shear test specimen and die.

C.   Flyer Plate Tests

Flyer plate tests are used for quantitatively measuring the adhesion of plated coatings under dynamic loading conditions. The principle of this kind of test is to create a compressive shock wave in a sample in such a way that the wave travels from the substrate to the coating perpendicularly to the outer surface. This wave is then reflected at the surface as a tensile shock wave propagating from the coating to the substrate. This tensile wave produces detachment of the coating if the peak stress value exceeds the adhesion of the coating (19). The test consists of utilizing magnetic repulsion to accelerate thin, flat metal flyer plates against the substrates under test in a vacuum. The flyer plate test apparatus, originally developed for shock wave testing of materials, consists of two conductors, a ground plate, and a flyer plate, separated by a thin insulation film of plastic. The conductors are connected so that the current flowing through them produces a magnetic repulsive force that drives the thin flyer (0.2 to 1.02 mm) away

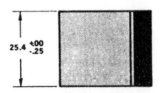

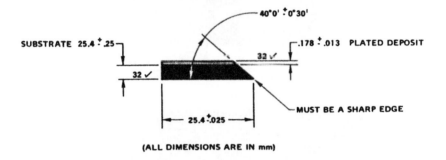

(ALL DIMENSIONS ARE IN mm)

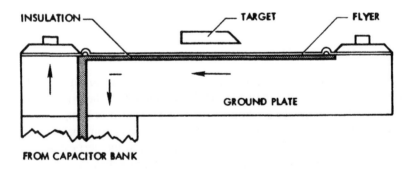

CROSS SECTION MAGNETIC FLYER ASSEMBLY

Figure 4: Flyer plate test specimen (top) and test apparatus (bottom).

from the relatively heavy ground plate and into the target specimen suspended above the flyer. The output of the flyer plate is determined by measuring the flyer velocity with a streaking camera. This velocity, coupled with metallographic cross sections of the specimens, provides the information needed to quantitatively compare different samples. Dynamic forces with amplitudes up to 100 kilobars and durations ranging from 100 to 500 ns have been obtained with a 14 kj capacitor bank. A test specimen schematic is shown in Figure 4 and additional details on the test can be found in references 20 and 21.

D. Peel Test

The peel test shown in Figure 5, was one of the early quantitative tests proposed for determining adhesion. Jacquet first used this method in 1934 to measure the adhesion of copper on nickel (22). A Jacquet type peel test has been used extensively since 1965 to determine the peel strength of plated plastics (23). For this test, an overlay of copper around 50 μm thick, is deposited on the plated plastic strip. Strips of metal, 25 mm (1 inch) wide, are then peeled normal to the surface. Peel strengths around 26 N/cm (15 lb/in) are commonly obtained. Klingenmaier and Dobrash have developed procedures for testing of various plated coatings on metallic substrates, and have used the peel test on production parts (24).

Figure 5: Peel test.

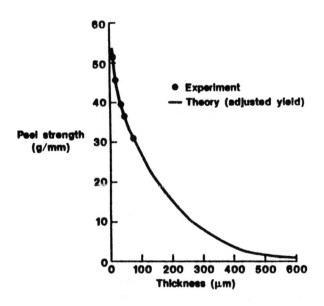

Figure 6: Comparison of the theoretical and experimental peel strengths as a function of Cr film thickness on a Si substrate. Adapted from reference 25.

Kim, et al. studied mechanical effects in the peel strength of a thin film both experimentally and theoretically (25). They reported that the adhesion strength measured by the peel test provides a practical adhesion value but did not represent the true interface adhesion strength. Factors which are of major importance in peel testing include: thickness, Young's modulus, yield strength, strain hardening coefficient of the film, compliance of the substrate and interface adhesion strength. A higher peel strength is obtained with thinner or more ductile films even though the true interface adhesion strength is the same. Figure 6 shows the comparison of the experimental peel strength for chromium films on a silicon substrate and theoretical values predicted by taking the strain-hardening factor and Young's modulus of the substrate into account. The theoretical model appears to be an excellent fit to experimental results and clearly indicates that most of the measured peel strength does not come from the true interface adhesion but from the plastic deformation of the film (25).

COMPARING ADHESION TEST RESULTS

Adhesion testing methods must duplicate the stresses to which the components are subjected in assembly and service. Very good adhesion in one test, does not necessarily mean that good adhesion will be obtained in another test because the failure mode could change drastically. This is shown in Table 2 which compares peel strength and ring shear data for nickel plated 6061 aluminum. With the peel test, the phosphoric acid anodizing activation process provided better adhesion than the double zincate process. However, in the ring shear test, the double zincate process provided greater than an order of magnitude higher adhesion. This shows that equating adhesion, which is a gross effect, to bonding or cleanliness may be very misleading. Failure of adhesion may be more related to fracture mechanisms than to bonding as mentioned earlier.

Table 2 -   Peel Strength and Ring Shear Data for Nickel Plated 6061 Aluminum

Preplate Treatment	Peel Strength N/cm(lb/in)	Ring Shear Strength MPa (psi)
Double zincate	107 (61), ref 24	200 (29,000), ref 26
Phosphoric acid, 40 V	140 (80), ref 24	17 (2,500), ref 27

TECHNIQUES FOR OBTAINING GOOD ADHESION

The purpose of this section is to provide a methodology for use with those substrates that are difficult to coat with an adherent electro-deposit. Table 3 lists a number of materials and classifies them according to ease of coating with adherent electrodeposits. The discussion that follows will be directed at those that require special treatment beyond routine cleaning and acid pickling to ensure adherence of the subsequent deposit.

The reason that some substrates are difficult to coat with adherent, deposits is that they have a thin naturally protective oxide film which reforms quite quickly when exposed to air. Therefore, even though a pickling operation might remove the oxide layer, it reforms before the part is immersed in the plating solution. One can remove such surface layers by sputter etching in vacuum but upon removal from the vacuum chamber the film reforms. For that matter, a gas monolayer can even form in one second at 10^{-5} Torr (9). Table 4 lists oxide thickness of some metals revealing the thinness of these troublesome layers.

A variety of techniques have been used to prepare difficult-to-plate

Table 3– Different Substrates Require Different Treatments to Provide Adherent Coatings

Easy to Plate	Special treatment	Very difficult
steel	stainless steels	titanium
copper	aluminum	molybdenum
brass	beryllium	tungsten
	magnesium	niobium
	plastics	tantalum
		glass

Table 4– Thickness of Oxide Films

Oxide	Thickness (Å)	Reference
Al_2O_3	18	28
Fe_2O_3	40	28
NiO	6,10	28,29
Ta_2O_5	16	30,31
300 Series StainlessSteel	20-100	30,31

substrates for coating. They include pickling in concentrated acids, mechanical roughening, intermediate strike coatings, displacement films, anodic oxidation, heating after plating, plasma/gas etching and physical vapor deposition using augmented energy (ion plating). Examples of each technique, itemized in Table 5, will be presented in the following sections.

A. Pickling in Concentrated Acids

Uranium is a good example to use to demonstrate how pickling in concentrated acids can help provide adhesion in some cases. If proper procedures are used, it is possible to obtain suitable mechanical adhesion between uranium and electrodeposited coatings. The most, successful techniques involve chemical pickling of the uranium in concentrated acid solution containing chloride ions (e.g., 500 g/l nickel chloride plus 340 ml/l nitric acid), followed by removal of the chloride reaction products in nitric acid before plating. This treatment does nothing more than provide a much

Table 5- A Variety of Techniques are Available for Preparing Difficult-to-Plate Substrates for Coating

Technique	Examples(discussed in text)
Pickling in concentrated acids	Etching uranium in nitric acid/nickel chloride solution
Mechanical roughening	Tantalum plated with nickel
Intermediate strike coatings	Wood's nickel strike; "glue" coatings on glass
Displacement films	Zinc films on aluminum and beryllium
Anodic oxidation	Phosphoric acid anodizing of aluminum
Heating after plating	Electroless nickel on aluminum; nickel on Zircaloy-2
Plasma/gas etching	Plating on plastics
Physical vapor deposition (ion plating)	Coatings on tungsten, molybdenum and titanium
Miscellaneous	Interface tailoring, oxide formation, partial pressure of gases, reactive ion mixing, phase-in deposition

increased surface area with many sites for mechanical interlocking or "interfingering" of the deposit. However, extremely good adherence can be obtained. Figure 7 shows the roughening and tunneling sites in etched uranium that provide the mechanical interlocking. Ring shear tests on parts receiving this type of treatment show failure in the coating rather than at the interface between the substrate and coating (32).

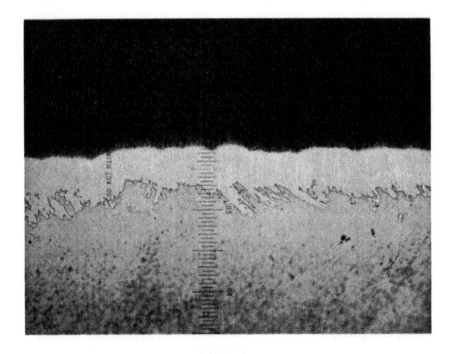

Figure 7: "Interfingering" developed in uranium as a result of etching in nickel chloride/nitric acid solution prior to nickel plating. Magnification is 300 ×.

B.   Mechanical Roughening

Tantalum is one of the most difficult metals to coat with an adherent electrodeposit. Results that have been reported previously are qualitative in nature, probably because quantitative data simply couldn't be obtained. Recent data show that by using mechanical roughening followed by anodic etching, reasonably good adhesion can be obtained. Adherent deposits of nickel, copper, and silver were obtained on tantalum when the tantalum was sandblasted and then anodically etched for 20 minutes at 200 A/m^2 in a methanol solution containing 2.5 v/o HCl and 2.5 v/o HF operated at 45°C. Depth of pitting as a result of the sandblasting/etching process was approximately 50-75 μm (2-3 mils). Peel strength data in Table 6 clearly show the importance of the mechanical roughening (sandblasting) part of the process. Without the mechanical roughening step, the subsequent anodic etch was extremely non-uniform and adhesion was considerably reduced.

Table 6-Peel Strength Data for Tantalum Electroplated With Nickel*

Preplating cycle	Current density in etch solution (A/m^2)	Average peel strength (lb/in)	(N/cm)
Sandblast, etch at 45 C for 20 min.	200	8.8	15.4
Sandblast, etch at 45 C for 20 min.	400	6.0	10.5
Scrub, etch at 45 C for 20 min.	400	3.8	6.7

*For comparison purposes. peel strength adhesion of nickel plated on aluminum is 100-200 N/cm (57-114 lb/in), reference 24.

C.   Intermediate Strike Coatings

The Wood's nickel strike (33) is an excellent example of use of an intermediate strike coating to improve adhesion. Typically used on stainless steels and nickel based alloys, a Wood's strike produced in a solution containing 240 g/l nickel chloride and 125 ml/l hydrochloric acid produces a thin, adherent deposit of nickel which serves as a base for subsequent coatings. Ring shear test results showing the value of using a Wood's nickel strike with 410 stainless are shown in Table 7. The only process that resulted in failure within the subsequent gold electrodeposit was that which included the Wood's nickel strike. Activation of stainless steel by immersion or cathodic activation in hydrochloric acid provided considerably inferior adhesion.

Table 7.   Influence of Various Activation Treatments on Ring Shear Adhesion of Gold Plated 410 Stainless Steel [A]

| | Shear Strength | | |
Treatment	MPa	psi	Location of Failure
Immersion in 6 percent (by weight) HCl	5	700	Gold-Stainless Steel Interface
Cathodic Treatment in 6 percent (by weight) HCl at 968A/m^2 for 2 min.	15	2,200	Gold-Stainless Steel Interface
Cathodic Treatment in 37 percent (by weight) HCl at 968A/m^2 for 2 min.	66	9,600	Gold-Stainless Steel Interface
Cathodic Treatment in Wood's Nickel Strike [B] at 108A/m^2 for 2 min.	152	22,000	Within Gold Deposit

(A) The gold was plated in a citrate solution at 32A/m^2. Stainless steel 410 contains 11.5 - 13.5 Cr and no Ni. From reference 34.

(B) The Wood's nickel strike solution contained 240 g/l nickel chloride and 120 ml/l HCl.

Table 8 shows the influence of current density used with the Wood's nickel strike on subsequent adhesion of gold or nickel deposits on stainless steel. When no nickel strike was used, failure occurred at the electrodeposit/substrate interface at very low strengths (5 MPa). When the nickel strike was used and overplated with gold, optimum adhesion was obtained when the current density in the Wood's solution was 108 A/m^2, or higher.

Prior to overplating with thick nickel in sulfamate solution, higher Wood's strike current densities were needed. Fairly strong bonds were obtained at current densities of 291 and 538 A/m^2, but maximum bond strength was not obtained unless the current density in the Wood's strike was 1080 A/m^2, or higher. The fact that a higher current density was

Table 8 - Influence of Wood's Nickel Strike Current Density on Ring Shear Adhesion of Gold or Nickel Plated AM363[A]

| Wood's Nickel Strike Current Density | | Ring Shear Bond Strength | | | |
| | | Gold | | Nickel | |
A/m^2	ASF	MPa	psi	MPa	psi
0	0	5	700	5	700
54	5	54	7,800	48	6,900
108	10	152	22,000	48	7,000
161	15	152	22,000	54	7,800
291	27	152	22,000	318	46,100
538	50	152	22,000	337	48,900
1080	100	152	22,000	488	70,700

(A) The cleaning/plating cycle consisted of anodic treatment at 323 A/m^2 in hot alkaline cleaner, rinsing, immersion in 18% (wgt) HCl for 2 minutes, rinsing, Wood's nickel striking (240 g/l nickel chloride,120 ml/l HCl) for 2 minutes, rinsing, and plating in either citrate gold solution at 32 A/m^2 or nickel sulfamate solution at 269 A/m^2. AM363 stainless steel contains 11.5 Cr, 4.5 Ni, 0.50 Ti, 0.04 C, 0.50 Mn, 1.0 Si and balance Fe. From reference 35.

required prior to nickel plating than prior to deposition of gold is attributed to the difference in shear strengths of the two electrodeposits, with nickel being much stronger.

With some stainless steels or nickel based alloys, a combination of anodic treatment in sulfuric acid followed by cathodic treatment in a Wood's strike may be necessary to insure a high degree of adhesion. Table 9 shows the benefit of using an anodic treatment in sulfuric acid solution prior to striking at 268 A/m^2 in a Wood's nickel solution when preparing 17-4 PH stainless steel for plating. The ring shear strength of parts given only a Wood's nickel strike was 195 MPa, whereas a combination of anodic treatment in sulfuric acid followed by the strike provided strengths of 472 MPa.

Table 9 - Ring Shear Data for Nickel Plated 17-4 PH Stainless Steel [A]

Cleaning/Activating Cycle	Ring Shear (MPa)	Strength (psi)
Clean [B], HCl Pickle, Wood's Nickel Strike at 268 A/m² for 5 min., Sulfamate Nickel Plate	195	28,200
Clean, HCL Pickle, Anodic Treat in 70 wt.% H_2SO_4 at 1070 A/m² for 3 min., Wood's Nickel Strike at 268 A/m² for 5 min., Sulfamate Nickel Plate	472	68,300

[A] The composition (in wt.%) of 17-4 PH stainless steel is 0.04 Carbon, 0.40 Manganese, 0.50 Silicon, 16.5 Chromium, 4.25 Nickel, 0.25 Iridium, 3.6 Copper and the remainder is Iron. From Reference 35.

[B] In all cases, the cleaning step included degreasing, then anodic and cathodic treatment in hot alkaline cleaner. The HCL pickle was 30 wt.%.

This is an unusual result since for most stainless steels, the ring shear test does not provide discrimination between Wood's nickel and anodic sulfuric acid treatment since failure typically occurs in the nickel deposit regardless of which procedure is used. For example, when AM363 stainless steel is plated with either nickel or nickel-cobalt, ring shear adhesion tests show no difference between Wood's strike activation and activation in anodic sulfuric acid followed by Wood's strike since all failures occur within the electrodeposited coatings (Table 10). By contrast, flyer plate tests which measure adhesion under highly dynamic conditions, show approximately a 50% improvement in bond strength when sulfuric acid treatment is used prior to Wood's striking (Table 10). This indicates that if electroplated stainless steel parts are to be used under conditions subjecting then to dynamic loading, use of anodic treatment in sulfuric acid prior to Wood's nickel striking is important.

Table 10 - Influence on Static and Dynamic Adhesion of Anodic Treatment in Sulfuric Acid Prior to Wood's Nickel Striking of AM363 Stainless Steel [A]

Activation Treatment	Electro- Deposit	Dynamic Adhesion Flyer Plate Spall [B] Threshold Velocity		Static Adhesion Ring Shear Strength	
		MP_a	psi	MP_a	psi
Wood's Strike	Nickel	4000	579,000	455	66,000
Anodic Sulfuric Plus Wood's Strike	Nickel	6000	868,000	455	66,000
Wood's Strike	Nickel-Cobalt	4480	648,000	559	81,000
Anodic Sulfuric Plus Wood's Strike	Nickel-Cobalt	6700	969,000	559	81,000

A. The complete preparation cycle included anodic cleaning in hot alkaline solution, rinsing, immersing in 18% (wgt) HCl at room temperature for one minute, rinsing, anodic treating in 70% (wgt) sulfuric acid at 1080 A/m^2 (100A/ft^2) for 3 minutes, rinsing, and then Wood's striking at 270 A/m^2 (25 A/ft^2) for 5 minutes prior to nickel or nickel-cobalt plating in sulfamate solution. In some cases, the anodic treatment in sulfuric acid was omitted as indicated above. From reference 35.

B. Spall is the separation of the plated deposit from the substrate due to the interaction of two rarefraction waves.

Some practitioners prefer to use the Wood's strike anodically and then cathodically to help promote adhesion. It's important to note that if this is done, a risk that is taken is that the Wood's solution can become contaminated during the anodic portion of the cycle. The best approach to use if one prefers to use the Wood's strike in this manner is two have two solutions, one for the anodic cycle and the other for the cathodic cycle. This way, all contamination that is introduced during the anodic cycle will not be re-plated out during the cathodic cycle, and interfere with adhesion.

Glass is another example of material which requires an intermediate coating to promote adhesion of any subsequent coating. The conventional technology used by platers for metallizing nonconductors with stannous chloride/palladium chloride activation followed by electroless deposition prior to final electroplating is not acceptable for glass when thick coatings are required. With this approach, deposits thicker than around 12.5 μm (0.5 mil) easily separate from a glass substrate.

The generally accepted criterion for adhesion between an oxide substrate such as glass and a metal film is that the metal must be oxygen active to react chemically with the oxide surface, forming an interfacial reaction zone (4). Intermediate oxide layers can be achieved by depositing an oxygen active metal (A) onto an oxide surface (BO), promoting the reaction A + BO → AO + B at the interface (9). Materials with large heats of oxide formation such as niobium, vanadium, chromium, and titanium are effective (Table 11). The higher the negative heat of formation, the higher the affinity for oxygen. Deposition of a thin layer (1000-2000 A) of one of these metals on glass via vacuum evaporation can then be followed by a further metal layer, adherent to the intermediate layer. This is the basis of several multilayer systems such as Ti-Au, Ti-Pd-Au, and Ti-Pt-Au which are used commercially (4,36). Deposition of one of these "glue" or "binder" layers by evaporation followed by a thin layer of copper without breaking vacuum produces an adherent base for thick electroplating. Using this approach, glass parts have been coated with adherent copper as thick as 1 mm (37).

With glass, the interfacial region can change with time. For example, the adhesion of silver films on glass measured 75 days after

Table 11 - Heat of Formation of Various Metal Oxides *

Oxide	Heat of formation, kcal/mol
Nb_2O_5	-463
V_2O_3	-290
Cr_2O_3	-270
TiO_2	-218
WO_3	-200
MoO_3	-180
Cu_2O	-40
Ag_2O	-7
Au_2O_3	+19

* Values taken from Handbook of Chemistry and Physics, Chemical Rubber Co., Cleveland, OH (1971).

deposition increased by a factor of 2.5 compared with the initial adhesion values (38). Similar results have been obtained with vacuum evaporated aluminum on glass (39). This increase in adhesion is attributed to the migration of oxygen to the interface and the formation of a more extensive reaction zone (3). This concept has been used in the development of the "composite film" metallizing technique, where a partial pressure of reactive gas is used in the initial stages of metal deposition to form a graded composition interfacial zone by reactive sputter deposition (40) and is discussed in more detail in the section on miscellaneous adhesion enhancement techniques. Improved adhesion after aging has also been noted with plated plastics (23).

D.   Displacement Films

Aluminum, beryllium and magnesium are good examples of substrates where use of a displacement film prior to final electrodeposition can provide excellent adherence. A zinc displacement films works well on all three of these metals although other displacement coatings such as stannates have proven effective with aluminum. The solution used for the zinc immersion process for aluminum contains essentially zinc oxide and sodium hydroxide. The oxide on the surface of the aluminum is dissolved by the sodium hydroxide, leaving the bare aluminum to take part in a chemical displacement reaction, whereby three zinc atoms are deposited for every two aluminum atoms which pass into solution (41). Good adhesion has been obtained using several different compositions, suggesting that adhesion may not be sensitive to the exact proportion of ZnO and NaOH.

Conical head tensile tests with 2024 and 7075 aluminum given a double zincate treatment prior to plating with nickel showed failure in the aluminum. Figure 8 is a cross section showing a 7075 aluminum sample after testing. The failure in the aluminum is clearly evident; no damage is seen in the nickel plating or at the interface between the nickel and the aluminum.

Often a copper cyanide strike is used directly after the zinc immersion step to protect the zinc film and provide a base for subsequent deposition. Table 12 compares use of such a strike on a number of aluminum alloys prior to electroless nickel deposition. Ring shear tests were run with a number of different alloys--1100, 2024, 5083, 6061 and 7075. Electroless nickel deposits 25 μm thick were overplated with 1.5 mm of copper to provide the necessary thickness needed for the ring shear test specimens. Tests results showed that, in general, bond strengths were higher for the as-deposited specimens when a copper strike was used. One exception was the 2024 alloy, which was not affected by lack of a copper strike. Most seriously affected by lack of the copper strike was the 6061

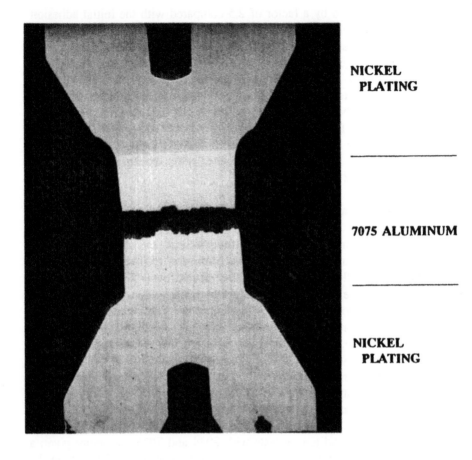

NICKEL
PLATING

7075 ALUMINUM

NICKEL
PLATING

Figure 8: Cross section of nickel plated 7075 aluminum conical head tensile specimen after testing. Magnification is 6 ×.

alloy (31 MPa without the strike versus 184 MPa with the strike). The 1100 and 7075 alloys exhibited lower bond strengths when a copper strike was not used, but upon heating to 149°C strengths were improved to levels comparable to those obtained with the copper strike (42).

With beryllium, very poor adhesion, less than 60 MPa was obtained when no zinc immersion treatment was used and also when the pH of the, zinc immersion solution was 9.3, or higher (Table 13). Specimens given a zincate treatment in solutions ranging in pH from 3.0 to 7.7 exhibited good adhesion; shear strengths ranged from 232 to 281 MPa (43).

Table 12 - Adhesion of Electroless Nickel Deposited on Various Aluminum Alloys[A]

Shear Strength [B] (MP_a)

Alloy	Zincate + Copper Strike + Electroless Nickel [C]		Zincate + Electro-Less Nickel	
	As-Deposited	Heat [D] Treatment	As-Deposited	Heat [D] Treatment
1100	99	90	68	108
2024-T351	221	246	251	204
5083-0	173	144	--	--
6061-T6	184	203	31	46
7075-T651	250	239	107	213

(A) The pre-plating cycle consisted of vapor degreasing, alkaline cleaning, rinsing, nitric acid pickling, rinsing, zincating for 30 sec., rinsing, nitric acid pickling, rinsing, zincating for 30 sec., rinsing, then either copper striking in cyanide solution or electroless nickel plating. The zincate solution contained 525 g/l hydroxide and 97.5 g/l zinc oxide. From reference 42.

(B) Each reported value is the average of five tests from one rod.

(C) All samples prepared with one proprietary electroless nickel solution, except for the 5083 alloy, which was plated using a different proprietary.

(D) Heat treatment was one hour at 149 C. Shear testing was done at room temperature.

Table 13 - Ring Shear Data for Nickel Plated Beryllium*

Process	pH	Shear Strength	
		MPa	psi
No zinc immersion treatment		0-51	0-7,400
Zinc immersion treatment	10.7	26	3,700
Zinc immersion treatment	9.3	60	8,700
Zinc immersion treatment	3.0	232	33,700
Zinc immersion treatment	3.2	241	35,000
Zinc immersion treatment	7.7	281	40,800

*Beryllium was S-200-E, 12.7 mm (0.5 in.) dia rod. The nickel plating solution contained 450 g/l nickel sulfamate, 40 g/l boric acid, <1.0 g/l nickel chloride. Current density was 268 A/m^2 (25 A/ft^2), pH 3.8-4.0, temperature 49°C, and anodes were SD nickel. From reference 43.

E.   Anodic Oxidation

Aluminum is the only metal that has been commercially prepared for reception of adherent electrodeposits by use of anodic oxidation. Although anodizing has been used as a pretreatment for plating on aluminum for over 50 years, the process isn't nearly as common as the zincate and stannate processes (44,45). The anodizing process, done in phosphoric acid solution, takes advantage of the oxidation characteristics of aluminum, and, in particular, the ability to form a porous anodic film under certain electrochemical conditions. Typical conditions include 15 to 50 volts for 10 minutes in a 36 wt% solution of phosphoric acid solution at 38°C (24).

The influence of varying the voltage during the phosphoric acid preplate treatment on the peel strength of several aluminum alloys is compared with that obtained with a double zincate treatment in Table 14. Different anodizing voltages were required for the three alloys given in Table 14 to obtain satisfactory adhesion. A substantially lower voltage (15 V) which was used to anodize the 3003 alloy produced peel strengths significantly higher than were obtained with the other alloys, 6061 and 7046. Both of these alloys showed increasing peel strengths with increasing voltage during anodizing. Although peel strengths comparable to those reached by zincating were obtained with the 6061 alloy by anodizing at 30 to 40 V, similar results were not obtained with the 7046 alloy even when anodizing as high as 50V. The 7046 alloy has a much higher alloy content which may contribute to the lower adhesion obtained by anodizing (24).

Although the anodizing process isn't nearly as popular as the zincating process for preparing aluminum for plating it has potential for future development. Low cost of chemicals, freedom from the use of cyanides, and lower waste treatment costs are definite advantages (44). The reason for the lack of popularity of this process apparently is due to the low adhesion on certain alloys, the high power required for some alloys and the variability of conditions required for different aluminum alloys.

F.   Heating After Plating

Heating after plating occasionally can improve adhesion. In some cases the heating is done at relatively low temperatures, e.g. 100 to 200°C, and in other cases the temperature can be very high, thus promoting rapid diffusion between the coating and substrate, e.g. electroless nickel on titanium heated at 850°C (46). Table 12, discussed earlier, provides data for various aluminum alloys coated with electroless nickel. Heating for one hour at 149°C was particularly effective in improving the bond strengths for 1100 and 7075 alloys which had been zincated and then directly plated with

Table 14 - Influence of Voltage During Phosphoric Acid Anodizing of Aluminum on Peel Strength of Subsequent Nickel Electrodeposits*

Alloy	Preplate Treatment	Peel Strength, N/cm (lb/in.) Range	Peel Strength, N/cm (lb/in.) Average Mean
3003	double zincate	173 to 245	196 (112)
	phosphoric acid anodize, 15V	183 to 227	198 (113)
6061	double zincate	70 to 210	107 (61)
	phosphoric acid anodize, 15 V	44 to 61	51 (30)
	phosphoric acid anodize 20V	70 to 88	79 (45)
	phosphoric acid anodize, 30 V	96 to 130	105 (60)
	phosphoric acid anodize, 40 V	131 to 149	140 (80)
7046	double zincate	96 to 210	156 (89)
	phosphoric acid anodize, 20 V	no adhesion	
	phosphoric acid anodize 30 V	17 to 35	26 (15)
	phosphoric acid anodize, 40 V	70 to 87	79 (45)
	phosphoric acid anodize, 50 V	70 to 123	96 (57)

*Peel test width was 25 mm. Thickness of nickel deposited in sulfamate solution was 160 μm. From reference 24.

electroless nickel (no intermediate copper strike). Ring shear bond strength of 1100 improved from 68 to 108 MPa as a result of heating and that of 7075 improved from 107 to 213 MPa.

Ring shear data for nickel plated Zircaloy-2 are shown in Table 15. The treatment cycle used to prepare the rods for plating consisted of etching in ammonium bifluoride solution. Neither this step nor any other chemical etching or other activation steps provided much higher adhesion in the as-deposited condition (47). To determine the influence of heating on adhesion, some rods were heated in vacuum at 700°C in both constrained and unconstrained conditions. The constraining was done by placing the specimens in a TZM molybdenum die with a 25 μm or less clearance and then heating. Since the coefficient of thermal expansion for the molybde-

Table 15 - Influence of heating on the ring shear strength of nickel-plated Zircaloy-2.[a]

	Ring Shear Strength, MPa		
Activation Treatment	As-deposited	700°C, 1h (uncon-strained)	700°C, 1 h (con-strained)[b]
Vapor degrease Cathodic alkaline clean Immersion in 15 g. NH₄FHF, 0.5 mL H₂SO₄ per litre, for 1 min at 22°C	16,25,15	38	140,292,224
Ni plate	. . .	. . .	. . .
Vapor degrease Cathodic alkaline clean Immersion in 45 g NH₄FHF per litre for 3 min at 22°C	31, 6	38, 12	235,234
Ni plate	. . .	. . .	. . .

[a] Each reported value is the average of at least two tests. From references 47 and 48.

[b] A TZM molybdenum ring was used to constrain the specimens during heating. Clearance between the specimens and the ring was 25 μm or less on the diameter.

num is lower than that of zirconium or nickel, it provided a stress calculated to be greater than 69 MPa on the electrodeposit as the assemblies were heated (48). The data in Table 15 clearly show a noticeable improvement as a result of heating for specimens which were constrained during heating. Bond strengths around 230 MPa were obtained for a number of specimens. By comparison, bond strengths in the as-deposited condition averaged 19 MPa. Heating in the unconstrained condition improved this to 29 MPa, still very low compared to the data obtained when constraint was used during heating.

G.   Plasma/Gas Etching

Chemical processes for plating on plastics are expensive, require rigorous chemical control, and present effluent treatment problems. Alternate processes that work well in preparing a variety of plastics for reception of adherent electrodeposits involve plasma or gas etching. Use of an oxygen, radio frequency (RF) glow discharge plasma treatment to condition ABS (acrylonitrile butadiene styrene) plastic followed by thin (1000 A) layers of nickel and copper applied by sputtering or electron beam evaporation provides a base for subsequent thick electrodeposition (49). The nickel layer helps establish a metal-plastic bond and the copper provides enhanced electrical conductivity for electroplating.   With the proper plasma voltage during the first step, adherence equal to that obtained by conventional chemical preplating (17.5 N/cm, and above) is obtained (Figure 9).

Figure 10 shows that the adhesion strength of vacuum deposited silver on modified polyethylene (PE) increases in the following order: untreated < argon plasma < oxygen plasma < nitrogen plasma treated PE (50). Argon plasma treatment of PE has little or no effect on the adhesion of vacuum deposited silver whereas oxygen and nitrogen plasmas improve adhesion of PE five and eightfold, respectively. Changes in the PE core levels after submonolayer deposition of Ag were interpreted as due to the formation of Ag-O-C and Ag-N-C species on oxygen and nitrogen plasma treated PE (50).

Another process that works well on ABS and ABS alloys including polycarbonate varieties involves ozone etching followed by alkali conditioning (51). The ozone gas "sees" and reacts uniformly with all exposed surfaces producing a very uniform etch. A significant advantage of the ozone process is that the reaction is diffusion controlled and, therefore, self-terminating. By contrast, chromic acid, used for aqueous etching, never stops reacting.

Vapor etching of plastics using sulfur trioxide is also gaining favor as a replacement for chromic acid etching prior to plating (52). During this

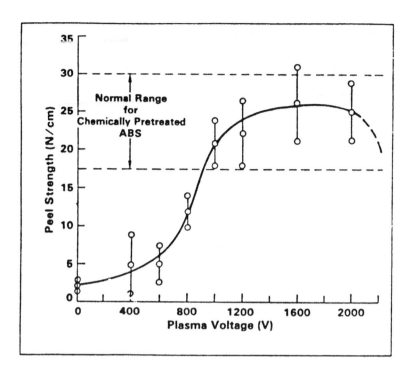

Figure 9: Peel strength vs. dc plasma voltage for ABS plastic. From reference 49. Reprinted with permission of The American Electroplaters & Surface Finishers Soc.

process, liquid sulfur trioxide is used and gasified on-site using a special generator system. At the end of the etch cycle, the gas is neutralized with ammonia before it is expelled from the chamber. The economics of this process are highly favorable when compared with chromic acid etching (53).

H.   Physical Vapor Deposition (Ion Plating)

A relatively new approach to providing adherence of electroplated coatings on difficult-to-plate substrates is that of utilizing augmented energy physical vapor deposition techniques such as ion plating to provide an initial layer which can be used as a base for subsequent, adherent electrodeposition.

Physical vapor deposition (PVD) is a process whereby a material in bulk form is atomistically converted to a vapor phase in a vacuum and condensed on a substrate to form a deposit. PVD techniques fall into three

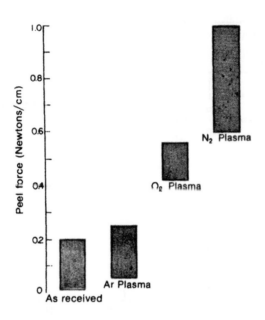

Figure 10: Comparison of adhesion strength of vacuum deposited silver on untreated and plasma treated polyethylene. From reference 50.

broad categories: evaporation, sputtering, and augmented energy. In evaporation, vapors are produced from a material located in a source which is heated by radiation, eddy currents, electron beam bombardment, etc. The process is usually carried out in vacuum (typically 10^{-5} to 10^{-6} Torr) so that evaporated atoms undergo an essentially collision-less line of sight transport prior to condensation on the substrate which is usually at ground potential (not biased). Sputtering involves positive gas ions (usually argon) produced in a glow discharge which bombard the target material thereby dislodging groups of atoms or single atoms which then pass into the vapor phase and deposit on the substrate. Augmented energy techniques have been employed recently to improve the energy of the deposition process. These include ion plating, thermionic plasma and formation of plasmas using hollow cathode discharge sources. Ion plating is a generic term applied to atomistic deposition processes in which the substrate is subjected to a flux of high energy ions sufficient to cause appreciable sputtering before and during film deposition. The ion bombardment is usually done in an inert gas discharge system similar to that used in sputter deposition, except that in ion plating the substrate is made a sputtering cathode. The substrate is subjected to ion bombardment for a time sufficient to remove surface contaminants and barrier layers (sputter cleaning) before deposition of the film material. After the substrate surface is sputter cleaned, the film deposition is begun without interrupting the ion bombardment. Variations in the ion plating technique have given rise to terms such as vacuum ion plating, chemical ion plating, bias sputtering, and others which refer to specific environment, sources or techniques.

With PVD films, adhesion varies from poor to excellent (54-56). Sputtering provides much better adhesion than the generally weak adhesion obtained with evaporation. This is due to the higher energy of the deposition species in sputter deposition as compared to that of evaporated atoms (1 to 10 eV vs. 0.1 to 0.2 eV) (13,57). By comparison, electroplating usually develops only a few tenths of an electron volt (58). With ion plating, adhesion is excellent and better than with sputtering. Again, this is attributed to the highly energetic deposition ions (>100 eV) as well as to the in-situ substrate precleaning and the simultaneous deposition and substrate sputtering due to the continuous bombardment of the substrate during deposition (54). Another factor that should be considered is the temperature at which sputtering or ion plating is done. Typically, use of higher temperature during either of these processes can lead to improved adhesion. Industrial applications where the technologies of ion plating and electroplating have been combined to provide adherence on difficult-to-plate metals can be found in references 59-63.

As mentioned in Table 3, the refractory metals such as tungsten, molybdenum and titanium are among the most difficult to electroplate with adherent, functionally thick acceptable deposits. All of the techniques discussed in sections A through G have been tried with these metals with less than optimal results. Recent efforts have shown that use of augmented energy physical vapor deposition to provide an initial adherent coating and then electroplating over this to final thickness provides excellent adhesion.

Ring shear data for tungsten, molybdenum and titanium are presented in Table 16. The results clearly show that use of an augmented energy PVD process for the initial stage of the coating cycle provided extremely good adherence. In all cases, the adherence was considerably improved over that obtained without use of the initial PVD layer. For example, the wet chemical process for tungsten consisted of etching parts in a HF/HNO$_3$ concentrated solution, followed by anodic treatment in 300 g/l KOH. Adhesion with this process was only 48 MPa, whereas, with ion plating as the initial step, adhesion was 173 MPa. With molybdenum, the plating process included two firing operations at 1000°C in dry hydrogen and the resulting adhesion of the gold electrodeposit was strong enough to cause failure in the gold at 125 MPa. By comparison, use of magnetron ion plating to provide the initial coating (60,000 A of copper) followed by thick electroplated copper resulted in a bond that failed in the copper deposit at

Table 16- Ring Shear Data Show the Value of Combining PVD with Electroplating for Coating Difficult-to-Plate Metals

Metal	Ring Shear Adhesion (MPa)	
	Electroplating	PVD & Electroplating
Tungsten	48[a](Cu)[b]	173[c](Cu)
Molybdenum	125[d](Au)	216[e](Cu)
Titanium	145[f](Ni)	252[g](Cu)

a) This process included etching in 3 parts HF, 1 part HNO_3, and 4 parts H_2O for 5 min. at 22°C followed by anodic treatment (1076 A/m^2) in 300 g/l KOH at 50°C for 5 min. prior to plating.

b) Metal in parenthesis was that used for building up the thick ring (1.5mm) required for ring shear testing.

c) The magnetron ion plating process included sputter etching in vacuum, magnetron ion plating with 6μm of copper and then electroplating to final thickness. Base pressure of the system was 5×10^{-6} Pa (10^{-8} Torr), etch power was 0.5 watts/cm^2, and bias power was 0.078 watts/cm^2 (but tapered to zero after deposition of about 20,000Å of copper).

d) This process included degreasing in perchlorethylene, firing in dry hydrogen (<2ppm H_2O) for 10 min., immersing in a solution containing four parts NH_4OH (28%) and one part H_2O_2 (30%) for 8 to 10 seconds at room temperature, rinsing in distilled water, gold striking to deposit 0.15 to 0.63 mg/cm^2 (0.08 to 0.32 μm), rinsing in distilled water, firing in dry hydrogen at 1000°C for 10 min., and then electroplating to final thickness.

e) The magnetron ion plating process included sputter etching in vacuum, magnetron ion plating with 6μ of copper and then electroplating to final thickness. Base pressure of the system was 5×10^{-6} Pa (10^{-8} Torr), etch power was 0.5 watts/cm^2 and bias power was 0.078 watts/cm^2 throughout the coating run.

f) This process included abrasive blasting, cleaning in hot alkaline solution, pickling in HCl, bright dipping in a solution containing 10% by vol. of HF (70%), 1% HNO_3 and balance water, followed by anodic etching for 6 min. at 162A/m^2 in a 40°C solution containing 13% by vol HF (70%), 83% acetic acid and 4% water. Then 25 μm of Ni was plated in a sulfamate solution at 48°C. Specimens were heated at 480°C for 2 hours and then plated with approximately 1.5 mm of nickel.

g) This process included coating with 10 μm of copper by hot hollow cathode deposition and then electroplating to final thickness. Conditions for the etch cycle included a source power/rate of 10Å/sec, substrate voltage of 2 KV and pressure of 3×10^{-4} Torr. Conditions for the coating cycle were a source power/rate of 100 Å/sec, and a pressure of 3×10^{-4} Torr.

216 MPa. Lastly, the same type of result was obtained with titanium. The wet chemical process included an anodic etch in HF/acetic acid and a heating step at 480°C for 2 hours and this provided a bond with a shear strength of 145 MPa. Use of ion plating provided a bond with a shear strength of 252 MPa. Clearly, coupling augmented energy PVD processes with electroplating provides better adhesion than obtained with the wet processes and also eliminates : 1) the need for roughening the surface chemically or mechanically and 2) heating after coating.

I.   Other Adhesion Enhancement Techniques

This section is intended to be futuristic in nature, and will build on the previous discussion of combining PVD techniques and electroplating since this approach represents new technology. A wide variety of methods have been used to improve the adhesion of vapor deposited films on solid substrates with which they have no chemical affinity in the bulk (7). Since the value of using PVD techniques to provide an intermediary coating has already been demonstrated, some discussion on techniques for improving adhesion of these films on difficult-to-plate substrates could be of future value. Techniques to be discussed include interface tailoring, alloying surface layers with metals exhibiting a high negative free energy of oxide formation, use of partial pressure of various gases during deposition, reactive ion mixing and phase-in deposition.

a. Interface Tailoring: The most direct techniques employ elevated substrate temperatures to increase the mobility of atoms arriving on the substrate and thereby improving the quality of the surface coverage (7). Recently developed beam processing techniques, however, offer fresh possibilities of interface "tailoring" which provides a powerful means of manipulating interface atomic structure in order to benefit adhesion. Providing interface irradiation by utilizing energetic ions (100 keV to several megaelectron volts) has been used to penetrate through metal films on various substrates to produce "stitching" at the interface. Modest doses (10^{15} - 10^{16} ions cm^{-2}) have been shown to improve adhesion dramatically in systems such as Cu/Al_2O_3 or Au/Teflon in which no bonding could initially be found (7).

In terms of tailoring the interface layers to optimize their performance in bonding, the standard ultrahigh vacuum methods of molecular beam epitaxy would often work. However, it has recently been shown that non-ultrahigh vacuum techniques, including use of large area, low energy (500eV) beams of ions before or during deposition, can offer valuable interface control and produce adhesion superior to that obtained by other methods. Strong bonds which were thermally stable have been produced by this manner with copper on alumina and Teflon (7,64,65).

   b. Free Energy of Oxide Formation: The adherence of amorphous chemically vapor deposited (CVD) alumina on pure copper was considerably enhanced by alloying a thin surface layer with preevaporated Zr, Al, or Mn; i.e., with metals exhibiting a high negative free energy of formation of the oxide similar to the situation with glass discussed earlier in this chapter. Table 17 compares the free energies of formation of oxides of the alloying elements with adhesion of subsequently applied alumina films. This clearly shows that the oxygen affinity of the metal, expressed by the negative free energy of formation of the oxide matches the predicted behavior (66).

   However, these results are not valid for a non-oxidizing process such as sputtering at low temperatures. During the first step of chemical vapor deposition of alumina, the intermediate layer is partially or totally

Table 17- Adherence of Chemically Vapor Deposited Alumina Coatings on Copper Pre-Coated With Various Films by Vacuum Evaporation (1,2)

Metal	Free Energy of Formation of Oxide (kcal/gram-atom of oxygen)	Thickness of Metal (A)	Adherence(3) (kg/mm^2)
Ag	+ 5	200	Very weak
Ni	- 43	20,200,1000	Very weak
Sn	- 51	200	Very weak
Mn	- 78	200	Strong (2.8)
Al	- 114	50,100,200,1000	Strong (1.6-2.3)
Zr	- 114	200	Strong

1. From reference 66.

2. During vacuum evaporation the copper substrate was heated to 250C, and during subsequent CVD treatment the temperature reached 500C.

3. Determined by bonding a 0.6 cm diameter pin to the substrate and pulling in tension.

oxidized, which is not the case during sputtering (67). Adherence of sputtered alumina to copper is poor, however, use of a titanium intermediate layer enhances adhesion significantly. Also, use of an intermediate nickel layer (which has a lower thermal expansion coefficient than copper) prior to titanium deposition further enhances adhesion (67).

In a somewhat similar manner, the existing technology for deposition of adherent films of copper onto alumina utilizes an initial layer of chromium to provide an adherent metal/oxide film. Since this technology results in a number of difficulties when these films are used for transmission lines or are etched for clearance for via-holes, an alternate process which eliminated the chromium interface was developed for providing good adhesion. It was shown that formation of a copper aluminate spinel ($CuAl_2O_4$) as an intermediate layer between the copper film and an aluminum oxide substrate worked quite well (68). Adhesion, measured by a pull test, showed a 50% improvement with the spinel interface as compared with the chromium interface as shown in Table 18.

c. *Partial Pressure of Various Gases:* As shown in Table 11, chromium is one of the "glue" or "binder" layers that works well in providing adhesion on glass due to the fact that chromium is oxygen active and reacts chemically with the oxide surface. An adhesion-pressure dependency has been shown to exist with results strongly influenced by oxygen in the system. Figure 11 shows that backfilling the deposition chamber with oxygen at pressures greater than 6.67 MPa produced superior adhesion as measured by a scratch test. Adhesion for oxygen backfill pressures less than 6.67 MPa was considerably reduced as was influence of CO_2, CO, H_2, Ar and CH_4 regardless of backfill pressure. The adhesion

Table 18- Adhesion of Copper Films on Alumina and Sapphire(1)

System (2)	Pull Strength MPa	psi
Cu/alumina	0.6	93
Cu/Cr/alumina	3.1	>454
Cu/Cr/sapphire	4.6	667
Cu/spinel/sapphire	6.3	>907
Cu/spinel/alumina	6.9	1005

I. From reference 68.

2. The copper films were deposited by vacuum evaporation; the spinel films were formed by heating after copper deposition.

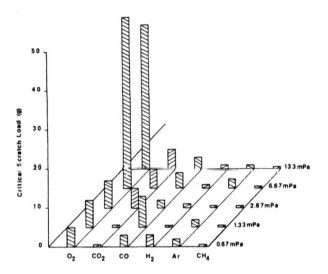

Figure 11:   Initial adhesion scratch loads of chromium on glass for hydrogen, argon, methane, oxygen, carbon monoxide, and carbon dioxide at various backfill system pressures. From reference 69.

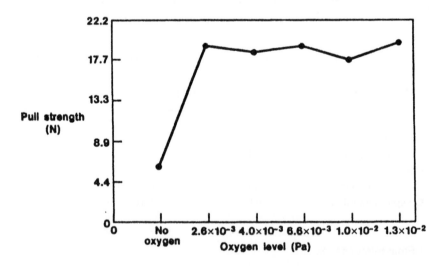

Figure 12:   Lead frame adhesion strengths and failure modes for various oxygen levels. Adapted from reference 70.

dependence of chromium corresponds to an oxygen/chromium impingement ratio of approximately 3, and it has been suggested that this ratio corresponds to the condition of critical condensation for a coherent chromium oxide adhesion layer (69). It was also noted that the critical scratch loads for all films tested were observed to increase with increasing time. This is attributed to the formation of an interfacial oxide layer of gradually increasing dimensions.

In a similar fashion, the influence of oxygen introduced into the evaporation system during the initial phase of chromium deposition on alumina substrates of hybrid microcircuits was shown to have a noticeable effect on adhesion. Figure 12 shows results from 90 degree pull tests of thermocompression bonded lead frames. With no oxygen in the system, very low pull strengths were obtained. The results for system pressure levels 2.67×10^{-3} through 1.33×10^{-2} Pa were excellent. Production yields of HMC's which had been as low as 45% exceeded 98% upon incorporation of oxygen backfill during chromium deposition (70). Another example is the sputter deposition of gold films on silica in a partial pressure of oxygen, which yield very adherent films without the use of an intermediate metal layer (71).

   d. Reactive Ion Mixing: Metal/polymer adhesion is becoming increasingly important to a variety of modern technologies such as the microelectronics industry where the development of flexible circuit boards, multilevel very large scale integrated (VLSI) interconnections, and advanced microelectronics packages are largely dependent on the integrity of the metal/polymer interface (72). Considerable attention has been devoted to polyimides (PIs) because of their excellent thermal, dielectric, and planarization properties. Recent work utilizing the unique capabilities of reactive ion mixing to modify the interfacial chemistry and adhesion of low reactivity metal/polyimide specimens has shown promising results. Implantation of $^{28}Si^+$ enhanced adhesion of Ni/PI specimens by a factor of 20 or greater (measured by the scratch test), such that the films were only removed as a result of failure in the PI substrates (Table 19). This adhesion increase was attributed to a combination of the substrate hardening, interfacial mixing (mechanical interlocking), interfacial grading and new chemical bonding characteristics (72).

   Use of $^{84}Kr^+$ implantation was also examined in order to isolate and evaluate the interfacial mixing (mechanical interlocking) and substrate hardening observed under $^{28}Si^+$ implantation without participating in any chemical bonding. The mechanical interlocking and substrate hardening induced by $^{84}Kr^+$ resulted in an adhesion increase of a factor of only 3, which was considerably smaller than that observed following $^{28}Si^+$ implantation (Table 19), indicating that the substrate hardening and mechanical interlocking had only a very small influence on the adhesion

Table 19 - Influence of $^{28}Si^+$ **and** $^{84}Kr^+$ **Implantation on Nickel/Polyimide Film Removal (a)**

Processing	Critical Force (N)	
	Ni/PI (adhesion)	PI substrate (toughness)
As-deposited	< 1(b)	10
1×10^{16} Si/cm^2	4	12
5×10^{16} Si/cm^2	8	15
1×10^{17} Si/cm^2	20 (c)	20
5×10^{15} Kr/cm^2	3	11
1×10^{16} Kr/cm^2	3	13
5×10^{16} Kr/cm^2	Ni film destroyed	22

a- The nickel film was 300 angstroms thick and deposited using electron beam evaporation. For details see references 72 and 73.

b- Evaluated by scratch test measurement. The lower limit with the test apparatus used was I N.

c- This measurement was limited by catastrophic substrate failure.

enhancement induced by $^{84}Kr^+$ implantation, and that the chemical bonding and interfacial grading produced by the $^{28}Si^+$ were primarily responsible for the large adhesion increases observed (73).

e. Phase-in Deposition: Phase-in deposition is a method widely employed in the electronic industry to provide improved adhesion for physically vapor deposited (PVD) films (74). As illustrated in Figure 13, this technique results in a physically intermixed layer in between two dissimilar metal layers. This intermixed layer consists of regions of different composition from 0% to 100% and often plays an important role in determining adhesion and overall physical and electrical properties of thin-film devices. For example, co-evaporation of Cr and Cu has been found to form a fine mix of Cr and Cu grains along the interface and this results in a very high adhesion strength which is superior to that obtained with Cr alone (74).

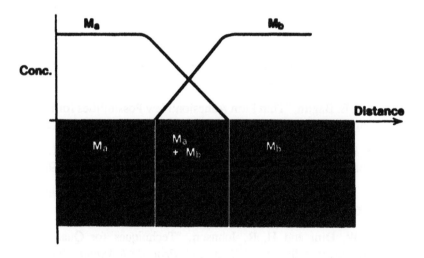

Figure 13: Schematic diagram of an electron beam evaporation system and the structure expected by phase-in deposition. Adapted from reference 74.

REFERENCES

1.   D. Davies and J. A. Whittaker, "Methods of Testing the Adhesion of Metal Coatings", *Metallurgical Reviews*, Review #112, 12, 15 (1967).

2.   *ASTM D907-70*, "Definition of Terms Relating to Adhesion", American Society for Testing and Materials.

3.   D. M. Mattox, "Thin Film Adhesion and Adhesive Failure-A Perspective", *Adhesion Measurement of Thin Films, Thick Films, and Bulk Coatings, ASTM STP 640*, K.L. Mittal, Editor, American Society for Testing and Materials, 54 (1978).

4.   D. M. Mattox, "Thin Film Metallization of Oxides in Microelectronics", *Thin Solid Films*, 18, 173 (1973).

5.   D. M. Mattox, "Fundamentals of Ion Plating", *J. Vac. Sci. Technol.*, 10, 47 (Jan/Feb 1973).

6.   E. C. Felder, S. Nakahara and R. Weil, "Effect of Substrate Surface Conditions on the Microstructure of Nickel Electrodeposits", *Thin Solid Films*, 84, 197 (1981).

7.   J. E. E. Baglin, "Thin Film Adhesion:New Possibilities for Interface Engineering", *Materials Science and Engineering*, B1, 1 (1988).

8.   B. K. Gupta, "Friction and Wear Behavior of Ion Plated Lead-Tin Coatings", *J. Vac. Sci. Technol.*, A5(3), 358 (May/June 1987).

9.   B. N. Chapman, "Thin Film Adhesion", *J. Vac. Sci. Technol.*, 11 106 (Jan/Feb 1974).

10.   J. W. Dini and H. R. Johnson, "Techniques for Quantitatively Measuring Adhesion of Coatings", *Proc. 18th Annual Conference*, Society of Vacuum Coaters, 27 (1975).

11.   *Properties of Electrodeposits: Their Measurement and Significance*, R. Sard, W. Leidheiser, Jr., and F. Ogburn, Editors, The Electrochemical Society, (1975).

12.   *Adhesion Measurement of Thin Films, Thick Films, and Bulk Coatings, ASTM STP 640*, K.L. Mittal, Editor, American Society for Testing and Materials, (1978).

13.   H. K. Pulker, A. J. Perry and R. Berger, "Adhesion", *Surface Technology*, 14, 25 (1981).

14.   *Testing of Metallic and Inorganic Coatings, ASTM STP 947*, W. B. Harding and G. A. DiBari, Editors, American Society for Testing and Materials, (1987).

15.   C. E. Moeller and F. T. Schuler, "Tensile Behavior of Electrodeposited Nickel and Copper Bond Interfaces", *ASM Metals Show*, Cleveland, Ohio (Oct 1972).

16.   J. W. Dini and H. R. Johnson, "Techniques for Quantitatively Measuring Adhesion of Coatings", *Metal Finishing*, 75, 42 (March 1977) and 75, 48 (April 1977).

17.   J. W. Dini and H. R. Johnson, "Adhesion Testing of Deposit-Substrate Combinations", *Adhesion Measurement of Thin Films, Thick Films, and Bulk Coatings, ASTM STP 640*, K.L. Mittal, Editor, American Society for Testing and Materials, 305 (1978).

18.   J. W. Dini, W. K. Kelley and H. R. Johnson, "Ring Shear Testing of Deposited Coatings", *Testing of Metallic and Inorganic Coatings, ASTM STP 947*, W. B. Harding and G. A. DiBari, Editors, American Society for Testing and Materials, 320 (1987).

19.   P. A. Steinmann and H. E. Hintermann, "A Review of the Mechanical Tests for Assessment of Thin-Film Adhesion", *J. Vac. Sci. Technol.*, A7 (3), 2267 (May/June 1989).

20.   J. W. Dini, H. R. Johnson and R. S. Jacobson, "Flyer Plate Techniques for Quantitatively Measuring the Adhesion of Plated Coatings Under Dynamic Conditions", *Properties of Electrodeposits, Their Measurement and Significance*, R. Sard, H. Leidheiser, Jr., and F. Ogburn, Editors, The Electrochemical Society, Pennington, NJ, 307 (1975).

21.   J. W. Dini and H. R. Johnson, "Flyer Plate Adhesion Tests for Copper and Nickel Plated A286 Stainless Steel", *Rev. Sci. Instrum.*, 46, 1706 (Dec 1975).

22.   P. H. Jacquet, "Adhesion of Electrolytic Copper Deposits", *Transactions of the Electrochemical Soc.*, 66, 393 (1934).

23.   E. B. Saubestre, L. J. Durney, J. Hajdu and E. Bastenbeck, "The Adhesion of Electrodeposits to Plastics", *Plating*, 52, 982 (1965).

24.   O. J. Klingenmaier and S. M. Dobrash, "Peel Test for Determining the Adhesion of Electrodeposits on Metallic Substrates", *Adhesion Measurement of Thin Films, Thick Films and Bulk Coatings, ASTM STP 640*, K. L. Mittal, Editor, American Society for Testing and Materials, 369 (1978).

25.   J. Kim, K. S. Kim and Y. H. Kim, "Mechanical Effects in Peel Adhesion Test", *J. Adhesion Sci. Technol.*, 3, No 3, 175 (1989).

26.   J. W. Dini, H. R. Johnson and J. R. Helms, "Ring Shear Test for Quantitatively Measuring Adhesion of Metal Deposits", *Electroplating and Metal Finishing*, 25, 5 (March 1972).

27.   J. W. Dini, previously unpublished information.

28.   H. J. Mathieu, M. Datta and D. Landolt, "Thickness of Natural Oxide Films Determined by AES and XPS With/Without Sputtering", *J. Vac. Sci. Technol.*, A3, (2), 331 (Mar/April 1985).

29.   M. R. Pinnel, H. G. Tompkins and D. E. Heath, "Oxidation of Nickel and Nickel-Gold Alloys in Air at 50-150 C", *J. Electrochem. Soc.*, 126, 1274 (1979).

30.   S. Kim and R. S. Williams, "Analysis of Chemical Bonding in TiC, TiN and TiO Using Second-Principles Band Structures From Photoemission Data", *J. Vac. Sci. Technol.*, A4, (3), 1603 (May/June 1986).

31.   J. R. Cahoon and R. Bandy, "Auger Electron Spectroscopic Studies on Oxide Films of Some Austenitic Stainless Steels", *Corrosion*, 38, 299 (June 1982).

32.   J. W. Dini and J. R. Helms, "Nickel-Plated Uranium: Bond Strength", *Plating* 61, 53 (1974).

33.   D. Wood, "A Simple Method of Plating Nickel on Stainless Steel", *Metal Industry*, 36, 330 (July 1938).

34.   J. W. Dini and J. R. Helms, "Electroplating Gold on Stainless Steel", *Plating* 57, 906 (1970).

35.   J. W. Dini and H. R. Johnson, "Plating on Stainless Steel Alloys", *Plating* 69, 63 (1982).

36.   A. T. English, K. L. Tai and P. A. Turner, "Electromigration in Conductor Stripes Under Pulsed DC Powering", *Appl. Physics Lett.*, 21, 397 (1972).

37.   W. C. Cowden, T. G. Beat, T. A. Wash and J. W. Dini, "Deposition of Adherent, Thick Copper Coatings on Glass", *Metallized Plastics 1, Fundamentals and Applied Aspects*, K. L. Mittal and J. R. Susko, Editors, Plenum Press, New York, 93 (1989).

38.   A. Kikuchi, S. Baba and A. Kinbara, "Measurement of the Adhesion of Silver Films to Glass Substrates", *Thin Solid Films*, 124, 343 (1985).

39.   P. Benjamin and C. Weaver, "The Adhesion of Evaporated Metal Films on Glass", *Proc. Royal Soc.*, A 261, 516 (1961).

40.   E. L. Hollar, F. N. Rebarchik and D. M. Mattox, "Composite Film Metallizing for Ceramics", *J. Electrochem. Soc.*, 117, 1461 (1970).

41.   D. S. Lashmore, "Immersion Deposit Pretreatments for Plating on Aluminum", *Plating & Surface Finishing*, 65, 44 (April 1978).

42.   J. W. Dini and H. R. Johnson, "Quantitative Adhesion Data for Electroless Nickel Deposited on Various Substrates", Proceedings Electroless Nickel Conference III (March 1983), *Products Finishing Magazine*, Cincinnati, Ohio.

43.   J. W. Dini and H. R. Johnson, "Joining Beryllium by Plating", *Plating & Surface Finishing*, 63, 41 (1976).

44.   D. S. Lashmore, "Electrodeposition on Anodized Aluminum: State of the Art", *Metal Finishing*, 78, 21 (April 1980).

45.   D. S. Lashmore,, "AES Research Project 41: Plating on Anodized Aluminum", *Plating & Surface Finishing*, 68, 48 (April 1981).

46.   G-Xi Lu and J-R. Liu, "Unlubricated Sliding Wear Behavior of Nickel Diffusion Coated Ti-6A1-4V", *Wear*, 121, 259 (1988).

47.   J. W. Dini, H. R. Johnson and A. Jones, "Plating on Zircaloy-2", *J. Less Common Metals*, 79, 261 (1981).

48.   J. W. Dini and H. R. Johnson, "Plating on Titanium and Zirconium", *Industrial Applications of Titanium and Zirconium: Third Conference, ASTM STP 830*, R. T. Webster and C. S. Young, Editors, American Society for Testing and Materials, 113 (1984).

49.   J. H. Lindsay and J. LaSala, "Vacuum Preplate Process for Plating on Acrylonitrile-Butadiene-Styrene,(ABS), *Plating & Surface Finishing*, 72, 54 (July 1985).

50.   L. J. Gerenser, "An X-ray Photoemission Spectroscopy Study of Chemical Interactions at Silver/Plasma Modified Polyethylene Interfaces: Correlations With Adhesion", *J. Vac. Sci. Technol.* A6 (5), 2897 (Sept/Oct 1988).

51.   J. M. Jobbins and P. Sopchak, "Chromic Acid Free-Etching", *Metal Finishing*, 83, 15 (April 1985).

52.   J. E. McCaskie, "Electroless Plating of Plastic Enclosures for EMI/RFI Shielding", *AES Second Electroless Plating Symposium* (Feb 1984), American Electroplaters and Surface Finishers Soc.

53.   J. H. Berg, "How to Use Vapor Etching for Effective EMI/RFI Shielding", *Electri-onics*, 32, No 10, 53 (Sept 1986).

54.   N. S. Platakis and L. Missel, "Wet and Vacuum Coating Processes", *Metal Finishing*, 76, 50 (July 1978).

55.   J. E. Varga and W. A. Bailey, "Evaporation, Sputtering and Ion Plating-Pros and Cons", *Solid State Technology*, 16, 79 (Dec 1973).

56.   P. R. Forant, "Vacuum Metallizing", *Metal Finishing*, 77, 17 (Nov. 1979).

57.   J. W. Dini, "Techniques for Coating Molybdenum", *Materials and Manufacturing Processes*, 4, No 3, 331 (1989).

58.   Bulletin SL-1002, "Sputtering as a Deposition Process", *Hohman Plating and Manufacturing*, Dayton, Ohio.

59.   B. C. Stupp, "Industrial Potential of Ion Plating and Sputtering", *Plating*, 61, 1090 (1974).

60.   J. W. Dini, "Electroplating and Vacuum Deposition: Complementary Coating Processes", *Plating & Surface Finishing*, 72, 48 (July 1985).

61.   T. G. Beat, W. K. Kelley and J. W. Dini, "Plating on Molybdenum", *Plating & Surface Finishing*, 75, 71 (Feb 1988)

62.   J. W. Dini and T. G. Beat, "Plating on Molybdenum-Part II", *AESF Aerospace Symposium*, (Jan 1989), American Electroplaters & Surface Finishers Soc.

63.   J. W. Dini, "Benefits From Combining Electroplating and Physical Vapor Deposition Technologies", *Proc. Surface Modification Technologies III*, T. S. Sudarshan and D. G. Bhat, Editors, The, Minerals, Metals and Materials Society (TMS), Warrendale, PA, 171 (1989).

64.   J. E. E. Baglin, A. G. Schrott, R. D. Thompson, K. N. Tu and A. Segmuller, "Ion Induced Adhesion Via Interfacial Compounds", *Nucl. Instrum. and Methods in Physics Research*, B19/20, 782 (1987).

65.   C.A. Chang, J. E. E. Baglin, A. G. Schrott and K. C. Lin, "Enhanced Cu-Teflon Adhesion by Presputtering Prior to the Cu Deposition", *Appl. Phys. Lett.*, 51 (2), 103 (1987).

66.   H. Schachner, R. Funk and H. Tannenberger, "Observations Concerning the Adherence of Amorphous CVD Alumina Coatings on Copper and Copper Alloys", *Proc. Fifth Int. Conf. Chemical Vapor Deposition*, J. M. Blocher, Jr., H. E. Hintermann and L. H. Hall, Editors, 485 (1975), The Electrochemical Soc.

67.   R. Jarvinen, T. Mantyla and P. Kettunen, "Improved Adhesion Between A Sputtered Alumina Coating and a Copper Substrate", *Thin Solid Films*, 114, 311 (1984).

68.   G. Katz, "Adhesion of Copper Films to Aluminum Oxide Using a Spinel Structure Interface", *Thin Solid Films*, 33, 99 (1976).

69.   R. S. Torkington and J. G. Vaughan, "Effect of Gas Species and Partial Pressure on Chromium Thin Film Adhesion", *J. Vac. Sci. Technol.*, A3 (3), 795 (May/June 1985).

70.   R. W. Pierce and J. G. Vaughan, "Using Oxygen Partial Pressure to Improve Chromium Thin Film Adhesion to Alumina Substrates", *IEEE Trans. on Components, Hybrids and Mfg. Technology*, Vol CHMT-6, No 2, 202 (June 1983).

71.   D. M. Mattox, "Influence of Oxygen on the Adherence of Gold to Oxide Substrates", *J. Appl. Physics*, 37, 3613 (1966).

72.   A. A. Galuska, "Adhesion Enhancement of Ni Films on Polyimide Using Ion Processing. I. $^{28}Si^+$ Implantation", *J. Vac. Sci. Technol.*, B8 (3), 470 (May/June 1990).

73.   A. A. Galuska, "Adhesion Enhancement of Ni Films on Polyimide Using Ion Processing. II. $^{84}Kr^+$ Implantation", *J. Vac. Sci. Technol.*, B8, (3), 482 (May/June 1990).

74.   J. Kim, S. B. Wen and D. Yee, "Coevaporation of Cr-Cu and Mo-Ag", *J. Vac. Sci. Technol.*, A6 (4), 2366 (July/Aug 1988).

4

DIFFUSION

INTRODUCTION

Diffusion is an irreversible and spontaneous reaction in a system to achieve equilibrium through the elimination of concentration gradients of the atomic species comprising the couple (1). Atoms or molecules within a material move to new sites and the net movement is from the direction of regions of low concentration in order to achieve homogeneity of the solution, which may be a liquid, solid or gas. Diffusion can result in degradation of the properties of a deposit, particularly at the basis metal interface. Examples of failures of electrodeposited components resulting from interdiffusion are numerous. In decorative applications, diffusion of the underlying metal to the surface can result in degradation of appearance (1). Copper plated zinc die castings often fail after exposure to elevated temperatures from the interdiffusion of the copper and zinc which results in the formation of a brittle intermetallic compound at or near the interface between the basis zinc metal and copper electrodeposit (1-3). In the case of gold plated contacts used in electronics applications, underlying copper can diffuse through the thin gold deposit to the surface where it oxidizes, and can significantly increase contact resistance. For example, one month of exposure at 250°C is sufficient to allow copper to diffuse extensively through 2.5 µm of gold overplate and at 500°C only three days are required for similar penetration of 25 µm of gold (4). Solderability of matte tin deposits over nickel stored at room temperature or slightly above deteriorates after a relatively short time due to the fast growth of a metastable $NiSn_3$ intermetallic compound (5). Intermetallics formed between electroless nickel and tin as a result of heating at 180-220°C have different densities which result in pores within the films (6).

GOOD ASPECTS OF DIFFUSION

Diffusion is not always bad and there are many instances where it is essential. For example, diffusion coatings rely upon interdiffusion for providing homogenization, tin coated steel relies upon the formation of the intermetallic compound $FeSn_2$ which forms in the diffusion zone and provides improved corrosion resistance (1), diffusion can improve the adherence between a coating and its substrate as discussed in the chapter on adhesion,and diffusion is a very key element in the joining of parts by solid state welding as shown later in this chapter.

USING DIFFUSION TO PRODUCE ALLOY COATINGS

Another example of the good aspects of diffusion includes the deposition of alloys. There are a number of cases wherein alternate layers of various coatings were deposited and then an alloy created by heating the sandwich structure to encourage diffusion and thereby produce an alloy. An alloy of 80% nickel-20% chromium was produced by depositing alternate layers of nickel (18.8 µm) and chromium (6.2 µm), and then heating for 4 hours at 1000°C. This treatment completely interdiffused the nickel and chromium, developing a homogeneous alloy which resisted air oxidation at elevated temperatures much better than nickel or chromium alone (7). Brass deposits have been formed by electrodepositing separate copper and zinc layers which were then thermally diffused to form a homogeneous alloy. This alternative of using diffusion formed brass eliminates many of the disadvantages of the cyanide process normally used to deposit the alloy, and allows more control over the properties of the coating (8). Diffused nickel-cadmium is used as a corrosion preventive plate for jet engine parts which may operate up to 480°C (9,10). Nickel thickness is 5-10 µm, cadmium thickness is 2.5-7.5 µm, and the homogenization temperature is 370°C. Parts chromated after this treatment have withstood in excess of 4000 hours in salt spray with no red rust (11).

DIFFUSION MECHANISMS

Diffusion in solid solutions occurs either interstitially or by substitution. In interstitial solid solutions, the solute atoms move along the interstices between the solvent atoms without permanently displacing any of them (Figure 1). This movement is encouraged by the fact that the solute atoms are much smaller than the solvent atoms. Small solute atoms can enter the lattice interstitially and diffuse to the nearly empty interstitial sublattice without the aid of defects and with a relatively low activation enthalpy. Hydrogen moves about lattices interstitially and this is the reason that considerable diffusion can occur rapidly even at temperatures used in many cleaning, pickling and plating operations (12). Other examples of interstitial diffusion are C, N, and O as solutes in bcc metals such as Fe (13).

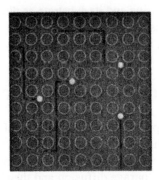

Figure 1: Interstitial diffusion. Adapted from reference 14.

Substitutional diffusion occurs by means of vacancies which are defects wherein atoms are missing from regular lattice positions. If an atom next to a vacancy is given a burst of energy it can move into the vacancy thereby leaving a new opening which can be filled similarly (Figure 2). Aside from the examples presented above which are representative of impurities in electrodeposits, essentially all electrodeposit-basis metal combinations are substitutional systems (1).

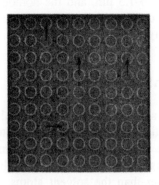

Figure 2: Substitutional diffusion. Adapted from reference 14.

There are three mechanisms of diffusion behavior; 1- bulk or lattice diffusion, 2-diffusion along defect paths and 3-ordered intermetallic phases. These are shown in Figure 3 and discussed in comprehensive fashion by Pinnel (4) from which the following was extracted.

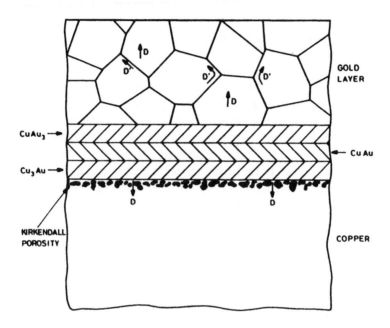

Figure 3: Schematic representation of the various diffusion related reactions induced by thermal aging of gold/copper structures. Bulk diffusion (coefficient D) corresponds to the transfer of copper into gold and gold into copper through the lattice. Defect path diffusion (coefficient D') involves transfer along lattice defects such as dislocations and twin or grain boundaries (only the latter is depicted here). Ordered Cu_3Au, CuAu, and $CuAu_3$ phases form at stoichiometric positions along the concentration gradient. Kirkendall porosity at the copper-gold interface illustrates the faster rate of diffusion of copper into gold as compared with that of gold into copper. From reference 4. Reprinted with permission of The Gold Bulletin.

Bulk or lattice diffusion involves vacancy-atom exchange. It is highly temperature dependent and temperatures approaching the melting point of the lower melting element are usually necessary for the reaction to near completion in a practical amount of time. As shown in Figure 3, where gold diffuses into copper as well as copper into gold, the lattice diffusion process is usually two-way. If the rates of diffusion are unequal, excess vacancies are left behind in the layer of the faster diffusing element, and these coalesce into a line of porosity known as Kirkendall porosity or voids (subsequently discussed).

Diffusion along defect paths has a lesser temperature dependence than lattice diffusion and usually becomes dominant at lower temperatures; for many systems this is room temperature. Defects such as grain boundaries, dislocations and twin boundaries can serve as rapid transport pipes through a metallic layer. This is particularly relevant for thin films where the techniques of application such as electrodeposition, sputtering and vapor deposition all provide very fine grained structures of high defect density. The rate constant defining this process has been found to be typically four to six orders of magnitude larger than the lattice diffusion coefficient at half the melting temperature in kelvins and below (4). Figure 4 is a plot of the log of the diffusivity versus 1/T where T is the temperature in kelvins. In the higher temperature mode (e.g., above 370°C), volume diffusion is predominant. In the lower ambient temperature mode, grain boundary diffusion becomes the primary diffusion mode as indicated by a change in slope for the diffusivity-temperature relationship (1,4,15).

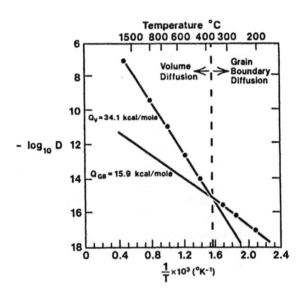

Figure 4: Plot of -$\log_{10}$ D vs 1/T showing the temperature dependence of the diffusion coefficients for volume and grain boundary diffusion for polycrystalline metals. Adapted from reference 1.

Ordered intermetallic phases between the coating and the substrate represent the third mechanism relating to interdiffusion (4). Figure 3 serves as an example for the gold/copper system where three ordered phases Cu_3Au, $CuAu$, and $CuAu_3$ can form and are stable below about 400°C. With this system the layer growth is relatively slow at these low temperatures and usually has no influence on the gold surface properties. However, it is important to note that ordered layers are often brittle and their formation can influence the mechanical integrity of the system (4).

KIRKENDALL VOIDS

As mentioned earlier, if interdiffusion between two metals is uneven since atoms of one metal may have a higher mobility and diffuse across the junction more rapidly than atoms of the other metal, this imbalance results in vacancies or voids at the interface between the two metals commonly referred to as Kirkendall voids or Kirkendall porosity shown in Figure 3 (16,17). With continued exposure to high temperature, the voids increase and coalesce causing porosity and loss of strength. This phenomenon has been observed in thin film couples of Al-Au (18), Au-Sn (19), Au-Pb (20) Au-Sn/Pb (21), Cu-Au (4, 22), Cu-Pt (23), and Pt-Ir (24) to mention a few. The Kirkendall effect is time and temperature dependent but with some systems can even occur at room temperature. For example, good adhesion of acid gold deposits to solder or solder to gold can be destroyed by heating at 120°C for as little as five minutes. Exposure of the bond to boiling water for 30 minutes will cause the plate to lift (21). Figure 5 shows the effect of time and temperature for copper plated with cobalt-hardened gold. The formation of Kirkendall voids is greatly accelerated by increased diffusion temperature and time. It is apparent that a line of porosity such as that in Figure 5 could lead to delamination of the gold layer (4).

Kirkendall porosity can be avoided by careful choice of the diffusing species. For example, Figure 6 shows Kirkendall voids that have caused partial separation of a platinum layer from a copper base. By contrast, Figure 7 shows no void formation between platinum and electrodeposited nickel after annealing for 8 hours at 700°C. In this case the diffusion coefficient of nickel is lower than that of copper due to the higher melting point of nickel (23). Annealing in a hot isostatic press (HIP) has also been used to suppress Kirkendall porosity formation (25).

One good aspect about the Kirkendall effect is that it can allow for the development of deliberately controlled porosity (26, 27). This technique has been proposed for electroforming parts requiring cooling channels. Figures 8 and 9 show this for tin-lead deposits which were overplated with gold and then heated to create the channel voids.

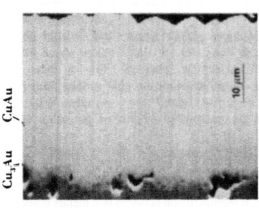

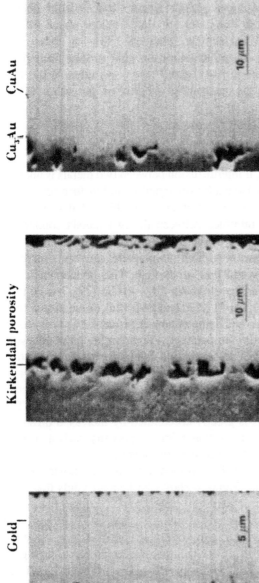

Figure 5: SEM back-scattered images showing the effects of aging time and temperature on the respective formation of layers of ordered phases and of Kirkendall porosity in cobalt-hardened or fine gold/copper systems. Ordering is slow below 25°C and not possible at or above about 400°C. The formation of Kirkendall voids is greatly accelerated by increased diffusion temperature and time. From reference 4. Reprinted with permission of The Gold Bulletin.

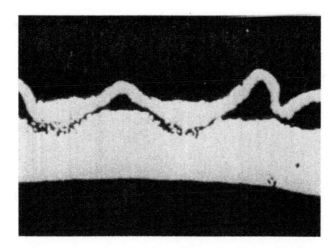

Figure 6:   Kirkendall voids formed between a platinum coating and a copper substrate (55×).   From reference 23.   Reprinted with permission of Platinum Metals Review.

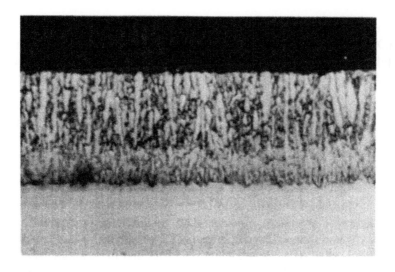

Figure 7:   This interface between a platinum coating and electrodeposited nickel shows no void formation after annealing for 8 hours at 700°C (125×). From reference 23.   Reprinted with permission of Platinum Metals Review.

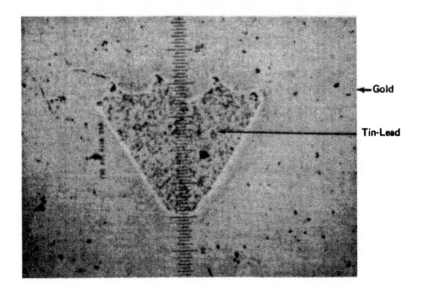

Figure 8:   Tin-lead deposit in gold.   Each division is equal to 3um (0.000125 in.).   Original magnification 300 ×.   From reference 26. Reprinted with permission of the American Electroplaters & Surface Finishers Soc.

DIFFUSION RATE

The rate of diffusion is influenced by a number of factors including: 1-the nature of the atoms, 2-temperature, 3-concentration gradients, 4-the nature of the lattice crystal structure, 5-grain size, 6-the amount of impurities, other alloying elements, vacancies and dislocations, and 7-the presence of cold work (14). Of these, the most important are temperature and the nature of the metals themselves.

1—Nature of the atoms-Overall diffusion rate is a measure of the combined effects of atom size, weight and charge on the nucleus. Generally the heavier and larger the atom, the slower it moves.

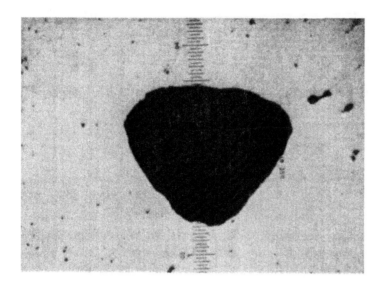

Figure 9: Tin-lead deposit in gold after heating at 500°C for 16 hours. Each division is equal to 3 µm (0.000125 in.). Original magnification 300 ×. From reference 26. Reprinted with permission of the American Electroplaters & Surface Finishers Soc.

2-Temperature-The noticeable influence of temperature on diffusion is approximated by the rule of thumb that the diffusion constant doubles for every 20°C (36F) increase in temperature (14,28). An example of the influence of temperature is shown in Figures 10a-d which are plots showing degradation of contact resistance for pure gold (0.5 µm) plated on copper as a function of heating time at various temperatures. The resistance of samples heated at 65°C shows no change even after 1000 hours of exposure (Figure 10a). However, reliability was seriously degraded in only 300 hours at 125°C and in 10 hours or less at temperatures of 200°C and higher (29).

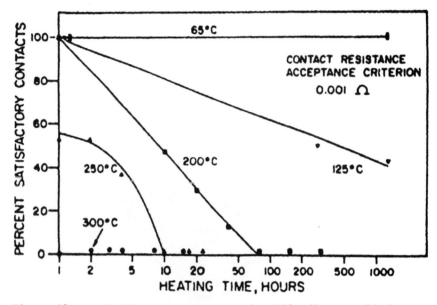

Figure 10a: Reliability vs. heating time for 0.02 mil pure gold plate on copper heated at 65, 125, 200, 250 and 300°C.   From reference 29. Reprinted with permission of the American Electroplaters & Surface Finishers Soc.

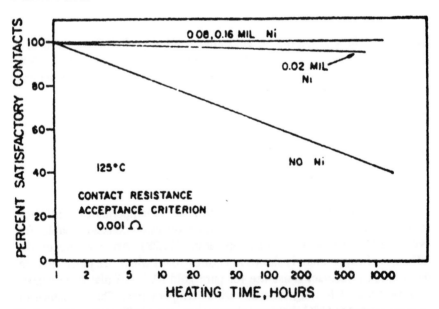

Figure 10b: Reliability vs. heating time for 0.02 mil pure gold plate on copper with no underplate, and with various thicknesses of nickel underplate. Heated at 125°C.

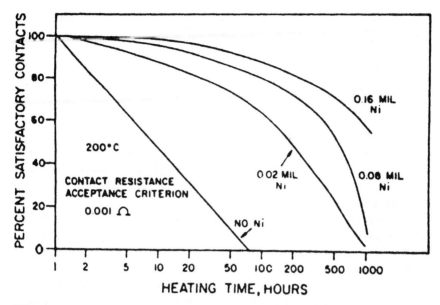

Figure 10c: Reliability vs. heating time for 0.02 mil pure gold plate on copper with no underplate, and with various thicknesses of nickel underplate. Heated at 200°C.

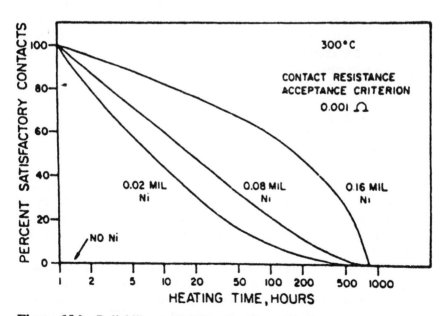

Figure 10d: Reliability vs. heating time for 0.02 mil pure gold plate on copper with no underplate, and with various thicknesses of nickel underplate. Heated at 300°C.

*3-Concentration gradients-*Since the driving force for diffusion is toward uniform composition, concentration gradients are important. The greater the difference in concentration, the greater the magnitude of the diffusion reaction.

*4-Lattice structure-*Some types of lattice structure are more conducive to diffusion than others. For example, with the more complicated structures such as hexagonal close packed, diffusion does not occur at the same rate in all directions as it does in cubic lattice systems.

*5-Grain size-*This is a less important factor since diffusion occurs much faster along grain boundaries than through the grains as discussed earlier. However, it is important to remember that the smaller the grain size the more the grain boundaries.

*6-Impurities and other alloying elements-*Alloying elements and/or impurities can noticeably influence diffusion. An example of this is shown in Table 1 which lists the maximum times that thick (5μm) gold and gold alloy deposits can be heated at various temperatures before they become unreliable from a contact resistance viewpoint. It is clearly evident that alloy gold deposits degrade more quickly than pure deposits, especially at high temperature (29).

*7-Cold work-*Diffusion occurs more rapidly when a metal has been cold worked, since dislocation densities are increased and grain size is reduced. Of importance from the viewpoint of electrodeposition is the fact than many electrodeposits often appear quite comparable to cold worked metals. For example, electroplated copper has exhibited behavior expected of 100% cold worked metal. More information on this is presented in the chapter on properties.

DIFFUSION BARRIERS

A. Introduction

An effective way to retard diffusion is to use a barrier plate. One of the classic examples of a coating as a diffusion barrier is the use of electrodeposited copper some 100 μm thick which serves as a complete and impervious barrier to carbon penetration in all commercial carburizing processes (30).

Certain metals are used as barriers which tend to block transport of the substrate metal into the noble metal overplate. For example, nickel and nickel alloys as a layer between copper and gold overplate are known to inhibit the diffusion of copper into the gold. This is shown very effectively in Figures 10a-d which are reliability curves for pure gold (0.5 μm) plate on copper, with (Figures 10b-d) and without a nickel underplate (Figure 10a).

These plots clearly show the effectiveness of nickel in preventing diffusion and also that the nickel is most effective as the thickness increases (29).

Table 1 - Effect of Alloy Content of Gold Plate on Reliability*

	Maximum Heating Time for 100% Relaible Contact (hr)		
Temp. (C)	Pure Gold Plate	0.1 % Co-gold	19 kt gold
65	>1000	>1000	50
125	500	500	2
200	300	2	-
300	25	-	-

* Deposits were 5 um (0.2 mil) thick; criterion of failure was 0.001 ohm. From reference 29.

B. Electronics Applications

When layers of copper or copper alloys and tin are deposited sequentially, a continuous barrier coating such as nickel should be interposed between them to resist the effects of aging. Table 2 clearly shows this for specimens aged at 95°C. With no nickel diffusion barrier between a bronze layer and tin or between a bronze/copper/tin sandwich, a brittle intermetallic layer containing 61% tin and 39% copper formed in 12 days at 95°C. After 90 days of exposure, growth of the intermetallic had increased and Kirkendall voids were formed. After 120 days of exposure, complete separation of the coating system from the substrate was obtained. With a nickel barrier layer of at least 0.5 μm thick between the tin and copper or copper alloy, no failure was obtained even after 240 hours of exposure at 95°C (31).

Copper-tin intermetallic compounds are also readily formed when tin bearing solder connections are made to copper surfaces. These compounds continue to grow during the life of solder connections and represent potentially weak surfaces. Use of a 1 μm thick nickel deposit between the phosphor bronze substrate and the solder provides and effective barrier. Long term strength at 150°C of 60Sn-40Pb solder connections formed between phosphor-bronze clip-on terminals and thin film terminations were markedly increased with the nickel diffusion barrier (Figure 11) (32). For other information on diffusion barriers for electronic applications see references 33-36.

Table 2 - Influence of a Nickel Barrier Between Copper or Copper Alloys and Tin*

Substrate	Coating	Results of Thermal Aging**
Al	Bronze***/5 µm Sn	12 days - brittle intermetallic layer containing 61% Sn-39% Cu
Al	Bronze/25 µm Sn	90 days - further growth of intermetallic layer and formation of Kirkendall voids
Al	Bronze/7.5 µm Cu/5 µm Sn	120 days - complete failure
Cu	Bronze/7.5 µm Cu/5 µm Sn	
Al	Bronze/2.5 µm Ni/5 µm Sn	
Al	Bronze/25 µm Ni/5 µm Sn	240 days - no failure
Cu	25 µm Ni/25 µm Sn	

*   From reference 31
**   All samples were aged in an oven at 95 C
***   Proprietary tin/bronze strike pretreatment; thickness was 0.5 to 1.0 µm, composition was 90 Cu/10 Sn.

C. Diffusion of Oxygen Through Silver

There is rapid diffusion of oxygen through silver at high temperatures (>350°C), and silver plated parts heated at these temperatures

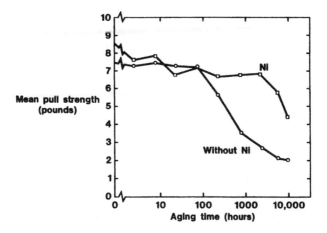

Figure 11: Aging results (150°C) for 60Sn-40Pb connections showing the influence of a nickel diffusion barrier 1.0 μm thick. Adapted from reference 32.

are likely to blister. The problem is overcome by applying a barrier layer which prevents the oxygen from passing through the deposit and oxidizing the underlying substrate. Recommendations include using 25 μm (1 mil) of copper or 1.3 μm (50 microinches) of gold. When gold is used, an air bake for 1 hour at 500°C is needed after the silver is applied to diffuse the gold into the substrate (37).

D. Nickel as a Diffusion Barrier for Brazing

Molybdenum and tungsten which have excellent high temperature properties are often brazed to iron for various applications. However, the direct brazing of iron to molybdenum or tungsten tends to cause exfoliation of the brazed joint in service due to formation of brittle intermetallic compounds such as Fe_7Mo_6 and Fe_7W_6. Nickel deposits 1.1 to 4.3 μm thick on the low carbon base metal restrain the formation of these brittle intermetallics thereby noticeably improving the mechanical properties and shear strength of the brazed joints (38).

DIFFUSION WELDING OR BONDING

This is a process that utilizes diffusion to make high integrity joints in a range of both similar and dissimilar metals. Clean, smooth surfaces are brought into intimate contact by a force insufficient to cause macroscopic deformation at an elevated temperature, usually in a vacuum or protective atmosphere. The problems, of inaccessible joints and unacceptable thermal cycles and resultant microstructures are mitigated, and distortion-free joints requiring no final machining may be produced (39).

The favorable features of diffusion welded joints include the following (40):

- Little or no change in physical or metallurgical properties
- No cast structures
- Minimization of recrystallization, grain growth, and precipitate dissolution
- Incorporation of heat treatment in the bonding cycle
- Multiple joints can be bonded simultaneously
- Excellent dimensional control
- Continuous gas-tight, extended area joints
- Minimization of weight and machining of finished product
- Preferred for dissimilar metal, cermet, and composite structures

Intermediate layers in the form of coatings or foils are often used to help promote joining and these coatings can be applied by electrodeposition. They are used for a variety of reasons including promoting plastic flow, providing clean surfaces, promoting diffusion, minimizing undesirable intermetallics, temporarily establishing eutectic melting to promote diffusion of base metals, minimizing Kirkendall porosity, reducing bonding temperature, reducing dwell time and scavenging undesirable elements (40,41).

Those coatings most frequently used include silver, nickel, copper and gold, with silver being used most often because of the low dissociation temperature of its oxide (42-44). Typical thickness range of electroplates used for diffusion welding is 12.5 to 35 μm but thicknesses as great as 125 μm have been used. Many different types of steel, aluminum, refractory metals, and beryllium have been joined with the aid of electroplated interfaces.

There are four critical process parameters common to all diffusion bonding techniques. They include temperature, pressure, time and surface condition/process atmosphere (40,44) and their interrelationship is shown in Figure 12. Bonding temperature is usually 1/2 to 2/3 the melting point of the lower melting point material in the joint. Use of elevated temperature serves to accelerate comingling of atoms at the joint interface and provides for metal softening which aids in surface deformation. The application of pressure serves the purpose of providing intimate contact of the surfaces to be joined and breaks up surface oxides thereby providing a clean surface for bonding. Dwell time at temperature is based on metallurgical and economic considerations. Sufficient time must be allowed to insure that surfaces are in intimate contact and some atom movement has occurred across the joint. However, too much atom movement can lead to voids within the joint or formation of brittle intermetallics.

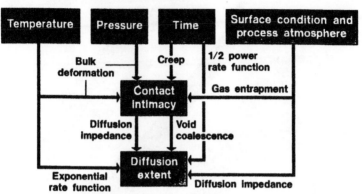

Figure 12: Effect and relationship of major diffusion bonding variables Adapted from reference 40.

The thickness of coatings has a noticeable influence on joint strength. Joints produced with a thin intermediate layer are subject to restraint of plastic flow during tensile loading; this results in triaxial tensile stresses that minimize the shear stress within the joint. The result is to prevent appreciable plastic deformation in the joint and to allow tensile strengths to be achieved that are many times larger than the bulk ultimate tensile strength of the intermediate layer material (45-47). The softer material is constrained between two high strength materials and the resulting triaxial stress state prevents a biaxial stress state which precludes deformation by shear. An example is that of Vascomax 250 maraging steel joints shown in Figure 13. The mechanical strength of these joints with a

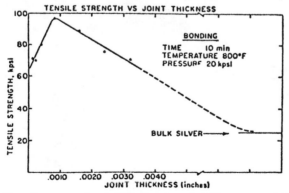

Figure 13: Tensile strength of diffusion bonded Vascomax 250 maraging steel coupons as a function of joint thickness. From reference 48. Reprinted with permission of The American Welding Society.

Figure 14a: 390 aluminum alloy valve body casting lapped and ready for plating. From reference 49. Reprinted with permission of The American Welding Society.

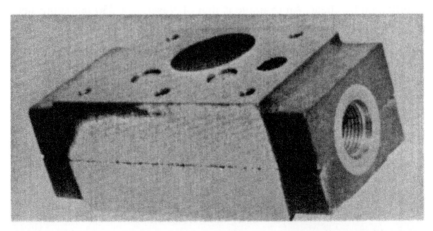

Figure 14b: 390 aluminum alloy casting after diffusion bonding. From reference 49. Reprinted with permission of The American Welding Society.

thin intermediate layer of silver goes through a maximum with decreasing joint thickness (48). For thick joints, the tensile strength is directly related to the bulk properties of the silver. As the joint thickness decreases, the tensile strength of the joint increases due to the restraints to plastic flow. For extremely thin intermediate layers, the problems of surface roughness and cleanliness start to hinder contact area and thus effectively reduce the tensile strength.

Applications include aluminum alloy hydraulic valve body castings (Figures 14a and 14b), aluminum and stainless steel tubing, hypersonic wind tunnel throat blocks (Figure 15) honeycomb stainless steel and aluminum, Inconel 600 screen, and copper cooling channels. In most of these cases, the materials being joined were metals difficult to coat adherently, e.g., stainless steel, aluminum, titanium, Zircaloy and nickel base superalloys (39).

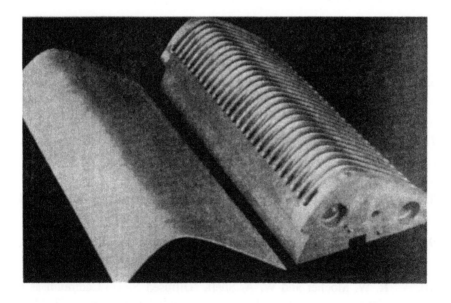

Figure 15:   Monel throat block and Be-Cu cover sheet which were subsequently plated with thin layers of gold and silver prior to diffusion bonding. From reference 50. Reprinted with permission of The American Welding Society.

REFERENCES

1.   E. L. Owen, "Interdiffusion", *Properties of Electrodeposits: Their Measurement and Significance*, R. Sard, H. Leidheiser, Jr., and F. Ogburn, Editors, The Electrochemical Society (1975).

2.   W. O. Allread, "Copper-Zinc Diffusion in Copper Plated Zinc Die Castings", *Plating 49*, 46 (1962).

3.   R. P. Sica and A. Cook, "Metallurgical and Chemical Considerations for Barrel Plating Zinc Die Castings to Withstand Prolonged 200°C Temperature and Possess High Corrosion Resistance While Maintaining Close Tolerances", *Proceedings SUR/FIN 90*, 103 (1990), American Electroplaters & Surface Finishers Soc.

4.   M. R. Pinnel, "Diffusion Related Behavior of Gold in Thin Film Systems", *Gold Bulletin*, 12, No. 2, 62 (April 1979).

5.   J. Haimovich and D. Kahn, "Metastable Nickel-Tin Intermetallic Compound in Tin-Based Coatings", *Proceedings SUR/FIN 90*, 689 (1990), American Electroplaters & Surface Finishers Soc.

6.   W. J. Tomlinson and H. G. Rhodes, "Kinetics of Intermetallic Compound Growth Between Nickel, Electroless Ni-P, Electroless Ni-B and Tin at 453 to 493 K", *Jour. Mater. Sci.*, 22, 1769 (1987).

7.   W. H. Safranek and G. R. Schaer, "Properties of Electrodeposits at Elevated Temperatures", *43rd Annual Technical Proceedings*, 105 (1956), American Electroplaters Society.

8.   D. A. Stout and R. J. Rife, "AES Investigation of Diffusion Formed Brass Coatings", *J. Vac. Sci. Technol.*, 20 (4), 1400 (April 1982).

9.   "Nickel-Cadmium Diffused", *Aerospace Material Specification, AMS 2416F*, Society of Automotive Engineers (1983).

10.   R. W. Moeller and W. A. Snell, "Diffused Nickel-Cadmium as a Corrosion Preventive Plate for Jet Engine Parts", *Plating 42*, 1537 (1955).

11.   J. F. Braden, "A Diffused Nickel Bond With a Post Plate", Proceedings Conference on Coatings for Corrosion Prevention, Phil. PA, *American Society for Metals* (1979).

12.   C. D. Beachem, "Mechanisms of Cracking of Hydrogen-Charged (Hydrogen-Embrittled) Aerospace Materials", *Proceedings AESF Aerospace Symposium* (Jan. 1989).

13.   A. S. Nowick, "Diffusion in Crystalline Metals: Atomic Mechanisms", *Encyclopedia of Materials Science and Engineering*, M. B. Bever, Editor, Pergamon Press 1180 (1986).

14.   "Diffusion and Surface Treatments", Chapter 13 in *Elements of Metallurgy*, R. S. Edelman, Editor, American Society for Metals (1963).

15.   L. Gianuzzi, H. W. Pickering and W. R. Bitler, "Factors Affecting Low Temperature Interdiffusion", *Proceedings AESF SUR/FIN 88*, (1988) American Electroplaters & Surface Finishers Soc.

16.   E. O. Kirkendall, "Diffusion of Zinc in Alpha Brass", *Trans. Metall. Soc. AIME*, 147, 104 (1942).

17.   A. D. Smigelskas and E. O. Kirkendall, "Zinc Diffusion in Alpha Brass", *Trans. Metall. Soc. AIME*, 171, 130 (1947).

18.   I. A. Blech and H. Sello, "Some New Aspects of Gold-Aluminum Bonds", *J. Electrochem. Soc.*, 113, 1052 (1966).

19.   S. Nakahara, "Microporosity in Thin Films", *Thin Solid Films*, 64, 149 (1979).

20.   J. R. Lloyd and S. Nakahara, "A Room Temperature Interdiffusion Study in a Gold/Lead Thin Film Couple", *Thin Solid Films*, 54, 207 (1978).

21.   B. Rothschild, "Solder Plating of Printed Wiring Systems", *AD 868621L*, (Sept. 1969).

22.   L. G. Feinstein and J. B. Bindell, "The Failure of Aged Cu-Au Thin Films by Kirkendall Porosity", *Thin Solid Films*, 62, 37 (1979).

23.   D. Ott and Ch. J. Raub, "Copper and Nickel Alloys Clad with Platinum and its Alloys", *Platinum Metals Review*, 31, (2), 64 (1987).

24.   B. Jones, M. W. Jones, D. W. Rhys, "Precious Metal Coatings for the Protection of Refractory Metals", *J. Inst. of Metals*, 100, 136 (1972).

25.   I. D. Choi, D. K. Matlock and D. L. Olson, "Creep Behavior of Nickel-Copper Laminate Composites With Controlled Composition Gradients", *Metallurgical Transactions A*, 21A, 2513 (Sept. 1990).

26.   H. R. Johnson and J. W. Dini, "Fabricating Closed Channels by Electroforming", *Plating 62*, 456 (1975).

27.   F. Aldinger, "Controlled Porosity by an Extreme Kirkendall Effect", *Acta Metallurgica*, 22, 923 (1974).

28.   *Elements of Physical Metallurgy*, A. G. Guy, Addison-Wesley Press, Cambridge, Mass. (1951).

29.   M. Antler, "Gold Plated Contacts: Effects of Heating on Reliability", *Plating 57*, 615 (1970).

30.   B. D. Whitley, P. C. Thornton and V. D. Scott, "Inhibition of Carburization by Gold Films", *Gold Bulletin*, 11 (2), 40 (April 1978).

31.   S. R. Schachameyer, T. R. Halmstad and G. R. Pearson, "Evaluation of Improved Reliability for Plated Aluminum Extrusions", *Plating and Surface Finishing*, 69, 50 (October 1982).

32.   H. N. Keller, "Solder Connections with a Ni Barrier", *IEEE Trans. on Components, Hybrids and Mfg. Technology*, CHMT-9, No. 4., 433 (December 1986).

33.   J. C. Turn and E. L. Owen, "Metallic Diffusion Barriers for the Copper-Electrodeposited Gold System", *Plating*, 61, 1015 (1974).

34.   M. R. Pinnel and J. E. Bennett, "Qualitative Observations on the Diffusion of Copper and Gold Through a Nickel Barrier", *Metallurgical Transactions A*, 7A, 629 (May 1976).

35.   D. R. Marx, W. R. Bitler and H. W. Pickering, "Metallic Barriers for Protection of Contacts in Electronic Circuits from Atmospheric Corrosion", *Plating and Surface Finishing*, 64, 69 (June 1977).

36.   D. R. Marx, W. R. Bitler and H. W. Pickering, "Metallic Barriers for Protection of Contacts in Electronic Circuits From Atmospheric Corrosion", *Atmospheric Factors Affecting the Corrosion of Engineering Metals*, ASTM STP 646, S. K. Coburn, Editor, American Society for Testing and Materials, 48 (1978).

37.   C. A. Kuster, "Silver Plating of Hot Gas Seals for High Temperature Applications in Rocket Engines", *Plating* 55, 573 (1968).

38.   T. Yoshida and H. Ohmura, "Effect of Nickel Plating on Fe-BCu-Mo and W", *Welding Journal*, 61, 363-s (Nov. 1982).

39.   J. W. Dini, "Joining by Plating", *Electrodeposition Technology, Theory and Practice*, L. T. Romankiw and D. R. Turner, Editors, The Electrochemical Society, Volume 87-17, 639 (1987).

40.   K. E. Meiners, "Diffusion Bonding of Specialty Structures", Proceedings 5th National SAMPE Technical Conference, Kiamesha Lake, NY (1973).

41.   C. L. Cline, "An Analytical and Experimental Study of Diffusion Bonding", *Welding Journal*, 45 (11), 481-s (1966).

42.   J. W. Dini, "Use of Electrodeposition to Provide Coatings for Solid State Bonding", *Welding Journal*, 61, 33 (Nov. 1982).

43.   J. W. Dini, W. K. Kelley, W. C. Cowden and E. M. Lopez, "Use of Electrodeposited Silver as an Aid in Diffusion Welding", *Welding Journal*, 63, 26-s (Jan. 1984).

44.   G. V. Alm, "Space Age Bonding Techniques-Part 1-Diffusion Bonding", *Mechancial Engineering*, 92, 24 (May 1970).

45.   H. J. Saxton, A. J. West and C. R. Barrett, "Deformation and Failure of Brazed Joints-Macroscopic Considerations", *Met. Trans.*, 2, 999 (1971).

46.   N. Bredz, "Investigation of Factors Determining the Tensile Strength of Brazed Joints", *Welding Journal*, 33 (11), 545-s (1954).

47.   W. G. Moffatt and H. Wulff, "Tensile Deformation and Fracture of Brazed Joints", *Welding Journal*, 42 (3), 115-s (1963).

48.   M. O'Brien, C. R. Rice and D. L. Olson, "High Strength Diffusion Welding of Silver Coated Base Metals", *Welding Journal*, 55, (1) 25 (1976).

49.   R. A. Morley and J. Caruso, "The Diffusion Welding of 390 Aluminum Alloy Hydraulic Valve Bodies", *Welding Journal*, 59 (8) 29 (1980).

50.   J. T. Niemann, R. P. Sopher and P. J. Rieppel, "Diffusion Bonding Below 1000°F", *Welding Journal*, 37, (8), 337-s (1958).

5
PROPERTIES

INTRODUCTION

The properties of electrodeposits are important for a broad spectrum of applications. Safranek summarizes these along with property data in his two texts on *The Properties of Electrodeposited Metals and Alloys* (1,2). The second volume of this set (published in 1986) contains property data from over 500 technical papers published since 1971 while the first volume (published in 1974 but now out of print) covers the previous years. Both of these are an invaluable help for anyone concerned with properties of deposits. Since they are so complete, and since properties are discussed throughout this book, this chapter will be relatively short.

A fundamental concern of materials science is the relationship between structure and properties and this is true for both bulk and coated materials (3). Hornbogen (4) divides the structural level of matter into six levels (Figure 1). The interactions which occur between these different levels of structure dictate the properties of engineering materials. These interactions may start just above one atomic spacing and extend over many grains. The structure-property relationships derived for thin films reflect this complex situation. A further complication is the fact that, in general, coatings are not deposited at equilibrium and contain high concentrations of lattice vacancies, dislocations, etc., which can vary from grain to grain (3).

Before going any further it is important to distinguish between mechanical and physical properties since they are often referred to improperly. Harold Read (5) made a clear distinction in 1960 and it is still applicable today. Those properties of metals and alloys which have to do

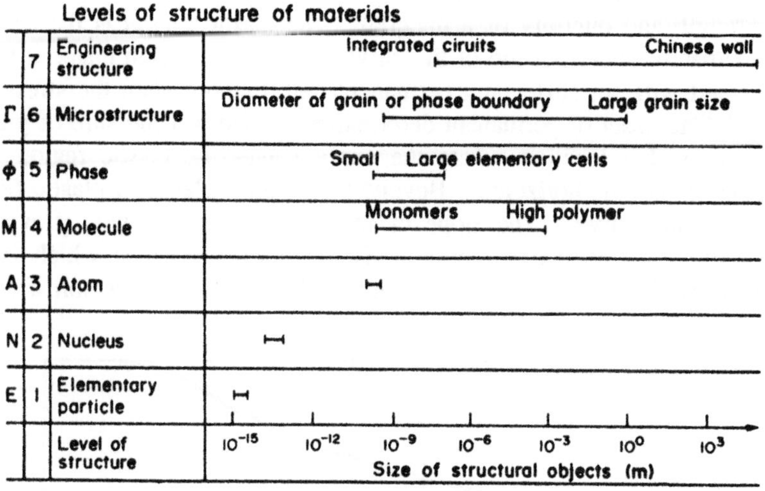

Levels of structure of materials

7	Engineering structure			Integrated ciruits		Chinese wall	
Γ 6	Microstructure	Diameter of grain or phase boundary			Large grain size		
φ 5	Phase		Small Large elementary cells				
M 4	Molecule		Monomers High polymer				
A 3	Atom			⊢⊣			
N 2	Nucleus	⊢⊣					
E 1	Elementary particle	⊢⊣					
	Level of structure	10^{-15} 10^{-12} 10^{-9} 10^{-6} 10^{-3} 10^{0} 10^{3}					

Size of structural objects (m)

Figure 1: The seven levels of structure suggested by Hornbogen (4). Reprinted with permission of Pergamon Press Ltd.

with strength, ductility, hardness, elastic modulus, and the like are properly called mechanical properties, not physical properties. The latter term is reserved for electrical conductivity, thermal conductivity, magnetic behavior, thermoelectric effects, density, melting point, lattice structure, etc. Perhaps the easiest way to divide non-chemical properties into their proper categories is simply to remember that properties which relate the deformation of a metal to a force which caused it are mechanical properties and all others are physical properties (5).

TENSILE PROPERTIES

The practical significance of the measurement of mechanical properties lies in the use of these data to predict the performance of a material in a specific type of application. Properties obtained from tensile testing are often used for engineering purposes.

A tensile stress-strain curve is constructed from load/elongation measurements made on a test specimen (6). Typically, original dimensions are used to calculate the stress based on load measurements and dimensions of the test specimen. This disregards any thinning or necking during testing and results in what is referred to as nominal or engineering stress. The terms true stress and true strain are used when actual dimensions during testing are used in the calculations (6).

    The shape of a stress-strain curve (Figure 2) is an indication of both the strength and ductility of a material. The elastic region is the early, approximately linear portion of the curve (7). In this region material that is stressed will not suffer any permanent deformation when the stress is relaxed. The onset of permanent deformation, which is a measure of yield strength, is that location where the curve leaves the elastic region by bending toward the horizontal. Beyond this is the inelastic or plastic flow region of the curve. The slope of the curve in each region provides information: in the elastic region it is the elastic modulus which is a measure of the material's stiffness and in the plastic flow region it is a

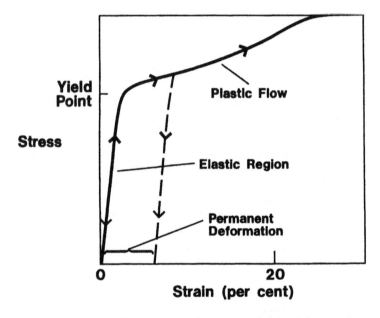

Figure 2: A representative stress-strain curve. Adapted from reference 7.

measure of work hardening since a steeper slope means more stress must be applied to create a given amount of deformation (7). Figure 3 shows the influence of strain rate on the strength and behavior of depleted uranium. Stress-strain curves are presented at strain rates of 5000 (dynamic) and 0.001 per second (static). The dynamic, or high strain rate curve reveals a higher yield point and, initially higher work hardening, followed by lower work hardening as the material thermally softens (7). Crack-free chromium is an example of an electrodeposit with a low strain hardening rate. With this low strain hardening rate, rapid localization of deformation occurs and this leads to early fracture and an increased wear rate unlike the behavior noted for conventional chromium deposits (8,9).

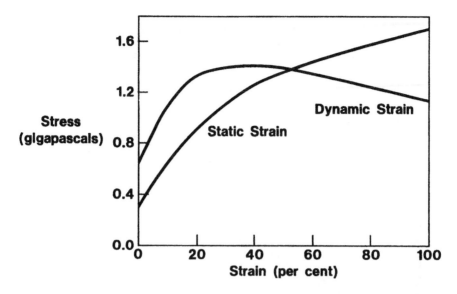

Figure 3: Stress-strain curves for depleted uranium at strain rates of 5000 (dynamic) and 0.001 per second (static). Adapted from reference 7.

The tensile strength of individual electrodeposited metals spans broad ranges and depends on the conditions adopted for electrodeposition (2). This is shown in Table 1 which compares strength of electrodeposits with their annealed, metallurgical counterparts of comparable purity. In a number of cases, the electrodeposit is two or three times as strong as the corresponding wrought metal. The maximum tensile strength for electrodeposited cobalt is more than four times the strength of annealed, wrought cobalt while some chromium deposits are nearly seven times as strong as cast or sintered chromium. A fine grain size is the primary reason for the higher strength of the electrodeposits as compared with their wrought counterparts (2). An example is electrodeposited gold containing 0.6 at% cobalt. This deposit has a hardness (VHN$_{10}$ = 190) about four times that of annealed bulk gold and this high hardness cannot be reproduced by standard metallurgical methods. The fine grain size (250–300Å) of the electrodeposited gold accounts for the observed high hardness. Other mechanisms such as solution hardening, precipitation hardening, strain hardening, and "voids" hardening account for only small alterations in the hardness of this coating (10).

The high strengths and hardnesses, high dislocation densities, fine grain structure, and response to heating obtained with electrodeposited metals are due to the existence of a strained condition similar to that found in cold worked metals. Figure 4, a plot of recrystallization temperature for electrodeposited pyrophosphate copper and wrought copper with various

Table 1: Strength and Ductility Data for Electrodeposited Metals and Their Wrought Counterparts (Ref. 2)

Metal	Plating Bath	Minimum Tensile Strength, psi	Maximum Tensile Strength, psi	Elongation, percent	Wrought Metal[a] Tensile Strength, psi	Elongation, percent
Aluminum	Anhydrous chloride-hydride-ether	11,000	31,000	2 to 26	13,000	35
Cadmium	Cyanide	-	10,000	-	10,300	50
Chromium	Chromic acid	14,000	80,000	<0.1	12,000	0
Cobalt	Sulfate-chloride	76,500	172,000	<1	37,000	-
Copper	Cyanide, fluoborate, or sulfate	25,000	93,000	3 to 35	50,000	45
Gold	Cyanide and cyanide citrate	18,000	30,000	22 to 45	19,000	45
Iron	Chloride, sulfate, or sulfamate	47,000	155,000	2 to 50	41,000	47
Lead	Fluoborate	2,000	2,250	50 to 53	2,550-3,000	42 to 50
Nickel[b]	Watts, and other type baths	50,000	152,000	5 to 35	46,000	30
Silver	Cyanide	34,000	48,000	12 to 19	23,000-27,000	50 to 55
Zinc	Sulfate	7,000	16,000	1 to 51	13,000	32

[a]Annealed, worked metal
[b]Data do not include values for nickel containing >0.005% sulfur

degrees of cold work suggests that the electrodeposited copper exhibits behavior expected of 100% cold worked material (11). The higher the percentage of cold work, or strained condition, the quicker the recrystallization behavior upon heating.   Other examples comparing electrodeposits with cold worked counterparts include copper deposited in acid solution containing thiourea and electrodeposited silver. Dislocation density of the copper deposits (12) was 3x 10^{11}/cm^2 compared to 2 x 10^{11}/cm^2 for cold worked copper (13) and the stored energy of cold worked silver reduced in cross section by 87% was of the same order of magnitude of that of electrodeposited silver (14).

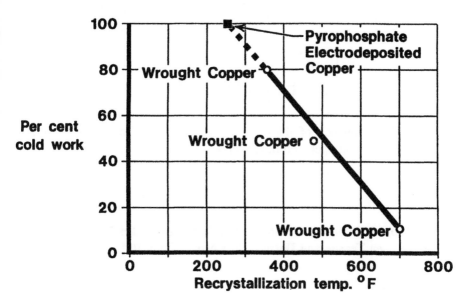

Figure 4:   Recrystallization temperature for various copper materials. Adapted from reference 11.

STRENGTH AND DUCTILITY OF THIN DEPOSITS

It's important to realize that tensile strength and ductility of thin deposits are very much influenced by the thickness of the test sample. Typically, tensile strength data are high for thin deposits and then decrease as a function of thickness before reaching some steady value while elongation data show the opposite. The effect on elongation is generally more pronounced than on tensile strength. Figure 5 and Table 2 show this effect for thick copper (0.2 to 3.0 mil) and nickel (5.5 to 144 mil) deposits respectively (15,16). This decrease in tensile strength and increase in

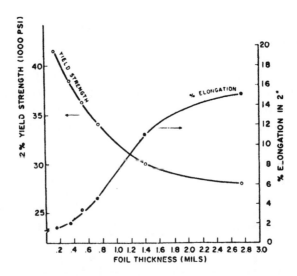

Figure 5: Influence of thickness on yield strength and elongation of electrodeposited copper. From reference 15.

Table 2: Influence of Thickness of Sulfamate Nickel Deposits on Tensile and Ductility Properties (From Ref. 16)

Thickness (mils)	Yield Strength (psi)	Tensile Strength (psi)	Elongation (%)	RA (%)
5.5	47 700	87 400	-	-
8.3	38 400	80 700	-	-
15	-	77 100	10.9	90.8
20	-	76 800	12.1	93.9
29	39 600	74 700	12.6	94.4
54	38 700	70 900	-	-
91	33 200	71 000	31.8	90.8
144	37 600	70 900	29.2	88.4

elongation with thinner deposits reflects a change in necking behavior whereby the thinner foils undergo less plastic flow at a given strain level (17).

Ductilites, as evidenced by percent elongation, are higher when the deposits are attached to their substrates since they cannot exhibit highly localized plastic deformation prior to fracture (18). The lower ductilities of the foils tested without their substrates are probably caused by the more severe local plastic deformation in the region where fracture subsequently occurs. This localized plastic deformation is called necking. When samples neck, this covers an appreciable portion of the cross sectional area so they break. Nickel electrodeposits (19) and electroless copper (20) have been found to exhibit severe necking when tested without their substrates.

In cases where it is possible, a better indicator of ductility is measurement of the reduction of area of the sample rather than elongation (17). Reduction of area is largely a measure of the inherent ultimate ductility of a material whereas elongation is largely a practical measure of stretching capability during forming; it is dependent on specimen shape and dimensions as well as on inherent ductility. Figure 6 shows three gold samples of varying thickness (1, 4, and 10 mils) after tensile testing. They all exhibit similar reduction in area values, e.g. greater than 95%, however, elongations varied from 5% for the 1 mil sample, to 11 % for the 4 mil sample to 23% for the 10 mil sample. The cross sections clearly show they were all equally ductile. This is a good example revealing that reduction in area measurements or metallographic cross section evaluation of a sample after fracture can give a better evaluation of the extent of necking strain.

Conventional tensile testing machines are used for samples with width-to-thickness ratios around 500. They are not suitable for thin foils with width-to-thickness ratios of 20, such as found on printed wiring boards (21). A new tensile testing machine developed under AESF Project 38 can

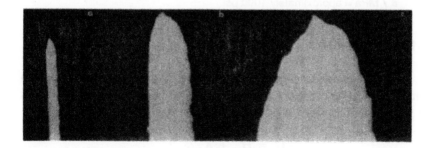

Figure 6: Gold deposits of varying thickness after tensile testing; a) 1 mil, 5% elongation, b) 4 mils, 11% elongation, and c) 10 mils, 23% elongation. As can be seen above, all exhibited greater than 95% reduction in area.

test samples as thin as 2000Å, which is only about 1000 atom layers (22,23).   Mechanical properties of thin nickel deposits tested on this machine are shown in Table 3.   Very thin deposits exhibited the highest yield strengths and this property decreased with increasing thickness.   The high yield strength of the thinnest deposit is probably due to surface pinning of dislocations.   The tensile strengths are seen to be relatively unaffected by the thickness of the deposits.   Elongation increased with increasing thickness because a larger portion of the gage length deformed plastically, with all deposits necking down to essentially the same thickness prior to fracture (23).

Table 3: Mechanical Properties of Thin Nickel Deposits[1]

Thickness (μm)	Young's Modulus (GPa)*	Yield Strength (MPA) #	Tensile Strength (MPa)	Percent Elongation
0.2	80 (3)	220 (5)	220 (5)	0
2.0	92 (1)	154 (3)	200 (5)	2.2 (0.2)
4.2	92 (0)	137 (7)	190 (6)	3.7 (0.2)
5.7	87 (1)	130 (11)	200 (7)	6.2 (0.2)
7.5	86 (3)	122 (8)	210 (13)	7.0 (0.1)
8.3	84 (2)	130 (8)	204 (4)	7.0 (0.3)
9.5	85 (2)	123 (7)	205 (5)	7.0 (0.2)

[1] From reference 23. Each value is the average of at least three measurements. The values in parenthesis are the variations on the individual measurements.
* 7 GPa = approximately 1 million psi
# 7 MPa = approximately 1000 psi

HALL-PETCH RELATIONSHIP

There have been several attempts to relate grain size of a metal with its mechanical properties.   One of these, the Hall-Petch (24) equation relates the grain size, d, with the hardness, H, of a metal:

$$H = H_o + K_H d^{-1/2}$$

The terms, H_o and K_H are experimental constants and are different for each metal. H_o is the value characteristic of dislocation blocking and is related to the friction stress. K_H takes account of the penetrability of the boundaries to moving dislocations and is related to the number of available slip systems (25). The equation has been found applicable to several polycrystalline materials as shown in Figure 7 (26), and also for electrodeposited iron (27), nickel (28,29) and chromium (30). Hall-Petch strengthening is shown in Figure 8 for electrodeposited nickel over a range of grain sizes from 12,500nm down to 12nm. A microhardness of approximately 700 kg/mm^2 was obtained with the smallest grain size (28). A Hall-Petch analysis has been used to help in understanding the occurrence of brittle cracking in chromium electrodeposits. An appreciable friction stress resistance to dislocation movement was shown to exist within the grain volumes of electrodeposited chromium and such friction strength strengthening for body centered cubic materials normally promotes brittleness (30). Hardness and grain size values for copper electrodeposits have also been analyzed using the Hall-Petch equation, but a better correlation was found using a semilogarithmic relationship (25).

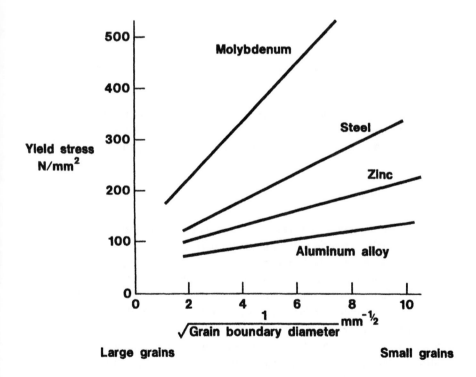

Figure 7: Hall-Petch relationship for a variety of metals. Adapted from reference 26.

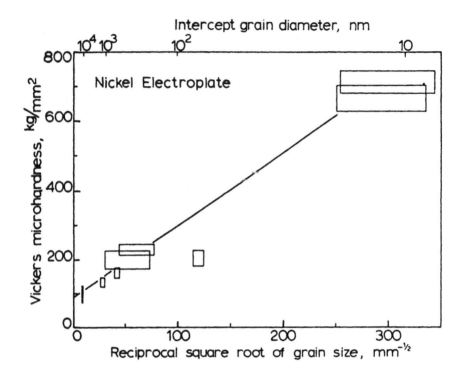

Figure 8: Microhardness dependence of electrodeposited nickel on reciprocal square root of grain size; rectangles define 95% confidence limits. From reference 28. Reprinted with permission of Pergamon Press Ltd.

SUPERPLASTICITY

The behavior of substances such as taffy and glass when they are heated to their softening point and gently pulled is noticeably plastic. These materials can be stretched to many times their original length and retain the new shape after they have cooled (31). A piece of metal under tension typically breaks before it reaches twice its original length, unless it is squeezed as it deforms (as it is in the drawing of wire) to counteract its tendency to "neck down" and break. Within the last twenty-five years, however, a number of alloys have been discovered that will behave like taffy or glass if certain significant steps are taken in their processing. This phenomenon is referred to as superplasticity and it offers the possibility of forming complicated shapes at high temperatures, while also increasing their room temperature strength, ductility and ability to be machined.

Superplasticity refers to large tensile elongations, typically 500%, that can be achieved in polycrystalline materials under certain conditions of strain rate and temperature. One of the main requirements for superplasticity is the presence of an ultrafine, equiaxed microstructure, typically 1 to 5 um diameter, that remains stable while being deformed at the superplastic temperature (usually around one-half the melting point) (32). This small grain size is typical of many electrodeposited metals, thus offering the exciting prospect that one could fabricate complex parts by combining preforming by electrodeposition and final forming of the internal structure by superplastic deformation. However, what isn't known for most electrodeposited metals is how stable these grain sizes are at temperatures around one-half their melting points. Two electrodeposited alloys which meet these requirements and, therefore, exhibit superplastic behavior are Cd-Zn (33) and Ni/40 to 60 percent Co (34,35). Elevated temperature ductility of Ni/45 to 56 percent Co is shown in Figure 9. A ductility peak occurs at 482 C (900 F), where Ni-Co exhibits superplastic behavior. A 280 percent elongation has been obtained over the central area of reduced test sections and significantly higher elongations may be possible since the process has not been optimized (34,35).

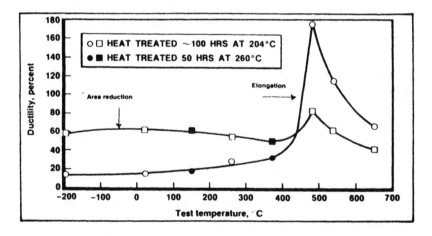

Figure 9: Average ductility of Ni-Co with 45 to 56 percent Co as a function of test temperature. From reference 34. Reprinted with permission of American Electroplaters & Surface Finishers Soc.

Although a number of alloys exhibit superplastic behavior, lead-tin (36,37), copper-nickel (38), and cadmium-tin (39) are three of particular interest since these can be deposited from aqueous solution. One technique that was used to produce superplastic lead-tin alloys was the deposition of alternate layers of 0.5 to 5 um thick (36), and this approach could be used to produce a variety of electrodeposited alloys.

The burgeoning field of electrodeposition of multilayer coatings by cyclic modulation of the cathodic current or potential during deposition (40) also offers promise for production of new superplastic alloys. Composition-modulated alloys (CMA) which have been produced by this process include Cu-Ni, Ag-Pd, Ni-NiP, Cu-Zn and Cu-Co. At present, no data on superplasticity of these alloys have been obtained, however, the room temperature tensile strength of CMA Ni-Cu alloys has been shown to exhibit values around three times that of nickel itself (41).

INFLUENCE OF IMPURITIES

Electrodeposited films contain various types of inclusions which typically originate from the following sources: 1-deliberately added impurities, i.e., organic or organometallic additives (addition agents), 2-metallic or nonmetallic particles for composite coatings, 3-intermediate cathodic products of complex metal ions, 4-hydroxides or hydroxides of a depositing metal, and 5-gas bubbles, for example, containing hydrogen (42). Figure 10 provides a pictorial illustration of these various types of inclusions. Much has been written on the influence of small amounts of inclusions on the appearance of deposits. However, very little information is available on their influence on properties of deposits. The purpose of this section is to provide examples showing how small amounts of impurities can noticeably affect properties.

With nickel, low current density deposits have higher impurity contents and this can affect stress and other properties. For example, Table 4 shows that for nickel sulfamate solution, hydrogen and sulfur contents are much higher for low current density deposits (54 A/m^2) than for those produced at higher current densities (43). Electrical resistance of electroformed nickel films shows a unique dependence on plating current density (Figure 11). Films deposited at a low current density of 120 A/m^2 show considerably lower residual resistance than high current density films over the temperature range of 4 to 40 K presumably due to codeposited impurities in the low current density deposits (44).

Small amounts of carbon in nickel and tin-lead electrodeposits can noticeably influence tensile strength. For example, increasing the carbon content of a sulfamate nickel electrodeposit from 28 to 68 ppm increased the tensile strength from 575 to 900 MPa, a noticeable increase in strength with a few ppm of the impurity (45). Similarly, with tin-lead, increasing the carbon content of the electrodeposit from 125 to 700 ppm increased the tensile strength from 29 to 41 MPa (46). Carbon also increases the strength

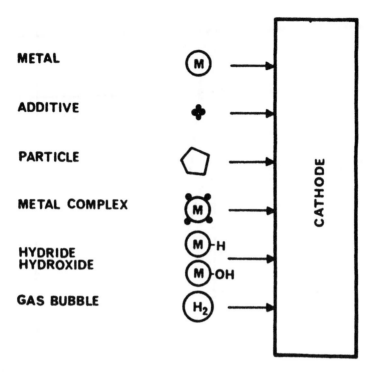

Figure 10: A pictorial representation of the various types of inclusions in electrodeposited films. From reference 42. Reprinted with permission of The Electrochemical Soc.

Table 4: Influence of Current Density in Nickel Sulfamate Solution on Impurity Content of Deposits (Ref 43).

Current Density		Impurity content (ppm)				
(A m⁻²)	(A ft⁻²)	C	H	O	N	S
54	5	70	10	44	8	30
323	30	80	3	28	8	8
538	50	60	4	32	8	6

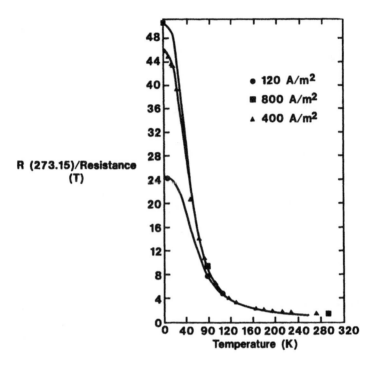

Figure 11: Resistance-temperature curves for electrodeposited nickel films approximately 20 um thick. Adapted from reference 44.

of cast nickel and nickel-cobalt alloys but the effect isn't as pronounced as that for electrodeposits. For example, increasing the carbon from 20 to 810 ppm in cast nickel increases the flow stress from 190 to 250 MPa (47).

Sulfur impurities can be harmful to nickel deposits which are intended for structural or high temperature usage. For example, small amounts of codeposited sulfur can noticeably influence notch sensitivity, hardness and high temperature embrittlement. Charpy tests, which are impact tests in which a center-notched specimen supported at both ends as a simple beam is broken by the impact of a rigid, falling pendulum, showed that deposits containing greater than 170 ppm of sulfur were highly notch sensitive (48,49). Figure 12 shows the results of testing specimens of two different thicknesses, 0.51 cm (0.200 in), and 0.19 cm (0.075 in). An increase in sulfur content is clearly shown to reduce the fracture resistance of electroformed nickel. Whereas thicker specimens (0.51 cm) displayed a steady decrease of impact energy with sulfur content, thinner specimens (0.19 cm) maintained roughly constant impact energy values up to 160 ppm. In this case, the thinner specimens were in a plane stress condition typified by shear fractures and relative insensitivity to sulfur content. In contrast, the

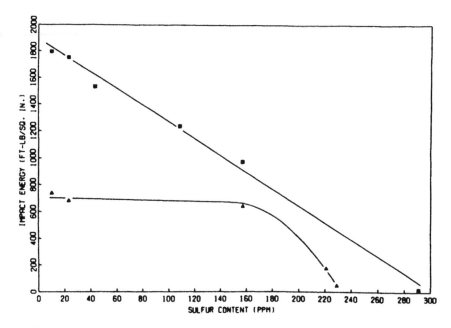

Figure 12: Influence of sulfur content on impact strength of electroformed sulfamate nickel. The squares are 0.200 in. (0.51 cm) thick Ni and the triangles are 0.075 in (0.19 cm) thick Ni. Adapted from reference 48.

plane strain condition (no strain in the direction perpendicular to the applied stress and crack length, reference 50) existing in thicker specimens led to higher triaxial tensile states and a significant sensitivity to sulfur content. Sulfur also has a direct influence on the hardness of electrodeposited nickel (Figure 13), therefore, if no other impurities are present in the deposit, hardness can be used as an indicator of sulfur content (48,49).

HIGH TEMPERATURE EMBRITTLEMENT OF NICKEL AND COPPER

Both nickel and copper electrodeposits undergo a ductile to brittle transition at high temperature. With nickel, reduction in area drops from greater than 90% at ambient to around 25% at a test temperature of 500 C (Figure 14, ref 51). This effect occurs at a much lower temperature for copper electrodeposits, e. g., 100 to 300 C depending on the conditions used for electrodeposition (Figure 15, ref 52).

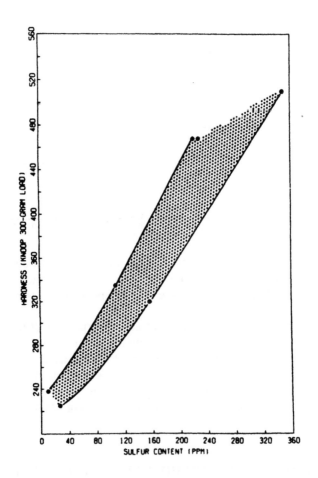

Figure 13: Influence of sulfur content on hardness of electroformed nickel. Adapted from reference 49.

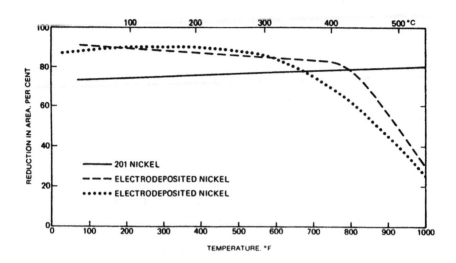

Figure 14: Influence of temperature on reduction in area of 201 nickel and electrodeposited sulfamate nickel. Adapted from reference 51.

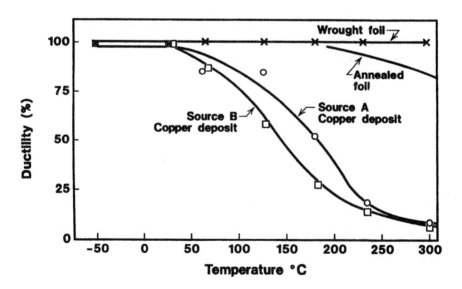

Figure 15: Influence of temperature on reduction in area for OFE (oxygen free electronic) copper and electrodeposited copper. Adapted from reference 52.

Electrodeposited nickel is quite pure, especially when compared with 201 wrought nickel which does not exhibit the ductile to brittle transition (Table 5 and Figure 14). The problem is that the electrodeposited nickel is too pure. Embrittlement occurs because of formation of brittle grain boundary films of nickel sulfide. Wrought 201 nickel doesn't exhibit the problem because it has sufficient manganese to preferentially combine with the sulfur and prevent it from becoming an embrittling agent. By codepositing a small amount of manganese with the nickel, the embrittling effect can be minimized. The amount of manganese needed to prevent embrittlement depends on the heat treatment temperature. The Mn:S ratio varies from 1:1 for 200 C treatments to 5:1 for 500 C treatments (51,53).

Embrittlement in electrodeposited copper is also probably due to grain boundary degradation stemming from the codeposition of impurities during electroplating. It's speculated that impurities modify the constitutive behavior or produce grain boundary embrittlement that leads to plastic instability and failure at small overall strains when compared with cast or wrought material of comparable grain size (54). At present the culprits have not been identified but two likely candidates are sulfur and oxygen. For example, cast high purity copper (99.999+%) is embrittled at high temperature when the sulfur content is greater than 4 ppm (55). Oxygen in cast copper has also been reported to cause embrittlement at high temperatures, either under tensile or creep conditions (56). This embrittlement is attributed to oxygen segregation to grain boundaries in the copper which promotes grain boundary decohesion and enhances intergranular failure. Both sulfur and oxygen can be present as impurities in electrodeposited copper.

OXYGEN IN CHROMIUM DEPOSITS

The relationship between the internal stress in chromium deposits and their oxygen content is shown in Figure 16. The broad band depicts the scatter observed in many hundreds of experiments (57). These variations are not unexpected because residual stress in any situation is related to the well known cracking of chromium deposits. The changes were achieved by changing the solution compositions at constant temperature (86 C) and current density (75 A/dm^2).

PHYSICALLY VAPOR DEPOSITED FILMS

With physically vapor deposited films, certain long term stability problems may be due to gas incorporation during deposition (58). In sputter

Table 5: Composition of 201 Nickel and Electrodeposited Sulfamate Nickel

Element	201 Nickel (ppm)	Electrodeposited Nickel[1] (ppm)
Copper	250 max	<100
Iron	400 max	<100
Manganese	3500 max	< 5
Silicon	3500 max	< 10
Carbon	93	50
Cobalt	4700	1000
Hydrogen	2	8
Oxygen	17	20
Nitrogen	6	6
Sulfur	12	10

[1] Composition of the nickel sulfamate plating soloution was 80 g/l nickel (as nickel sulfamate), <1.0 g/l nickel chloride, and 40 g/l boric acid. Wetting agent was used to reduce the surface tension to 35-40 dynes/cm. Current density was 268 A/m^2; pH, 3.8; and temperature, 49°C. Anodes were sulfur depolarized nickel.

deposition, up to several atomic percent of atoms of the sputtering gas can be incorporated into the deposited film and this gas can precipitate into bubbles or be released by heating (59-64). The incorporated gas can increase the stress and raise the annealing temperature of sputter deposited gold films (59). Argon incorporation up to 1.5 at. % is possible in TiC films and this causes compressive stresses of the order of 10^7 Pa. Such high stresses give rise to lattice distortion which affects the dislocation properties and thus the hardness of the films (60). Similar effects are found in electron beam evaporated films where residual gases, often released by heating during evaporation, are incorporated into the deposit and may cause property changes (64).

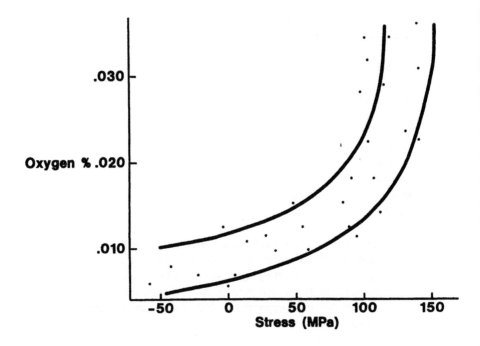

Figure 16: Influence of oxygen on stress in chromium electrodeposits produced at 86°C and 75 A/dm². Adapted from reference 57.

REFERENCES

1. W. H. Safranek, *The Properties of Electrodeposited Metals and Alloys*, American Elsevier Publishing Co., (1974)

2. W.H. Safranek, *The Properties of Electrodeposited Metals and Alloys*, Second Edition, American Electroplaters & Surface Finishers Soc., (1986)

3. D.S. Rickerby and S.J. Bull, "Engineering With Surface Coatings: The Role of Coating Microstructure", *Surface and Coatings Technology*, 39/40, 315 (1989)

4. E. Hornbogen, "On The Microstructure of Alloys", *Acta Metall.*, 32, 615 (1984)

5. H.J. Read, "The Metallurgical Aspects of Hydrogen Embrittlement in Metal Finishing", 47th Annual Technical Proceedings, *American Electroplaters Soc.*, 110 (1960)

6. "Testing for Materials Selection", *Advanced Materials & Processes*, 137, No 6, 5 (June 1990)

7. D. J. Sandstrom, "Armor Anti-Armor Materials by Design", *Los Alamos Science*, 17, 36 (Summer 1989)

8. D.T. Gawne and G.M.H. Lewis, "Strain Hardening of High Strength Steels", *Mater. Sci. Technol.*, 1, 128 (1985)

9. D.T. Gawne and U. Ma, "Friction and Wear of Chromium and Nickel Coatings", *Wear 129*, 123 (1989)

10. C.C. Lo, J.A. Augis and M. R. Pinnel, "Hardening Mechanisms of Hard Gold", *J. Appl. Phys. 50*, 6887 (1979)

11. D.E. Sherlin and L.K. Bjelland, "Relationship of Corner Cracking in Multilayer Board Holes to Pyrophosphate Copper Plate", *Circuit World*, 4, No 1, 22 (Oct 1977)

12. E.M. Hofer and H.E. Hintermann, "The Structure of Electrodeposited Copper Examined by X-ray Diffraction Techniques", *J. Electrochem. Soc.*, 112, 167 (1965)

13. R.E. Smallman and K.H. Westmacott, "Stacking Faults in Face-Centered Cubic Metals and Alloys", *Phil. Mag.*, 2, 669 (1957)

14.   W.F. Schottky and M.B. Bever, "On the Excess Energy of Electrolytically Deposited Silver", *Acta Met.*, 7, 199 (1959)

15.   T.I. Murphy, "The Structure and Properties of Electrodeposited Copper Foil", Finishing Highlights, 71, (Jan/Feb 1978)

16.   H.R. Johnson, J.W. Dini and R. E. Stoltz, "The Influence of Thickness, Temperature and Strain Rate on the Mechanical Properties of Sulfamate Nickel Electrodeposits", *Plating & Surface Finishing*, 66, 57 (March 1979)

17.   T.D. Dudderrar and F.B. Koch, "Mechanical Property Measurements on Electrodeposited Metal Foils", *Properties of Electrodeposits, Their Measurement and Significance*, R. Sard, H. Leidhelser, Jr., and F. Ogburn, Editors, The Electrochemical Society, Princeton, NJ 1975.

18.   P. Vatakhov and R. Weil, "The Effects of Substrate Attachment on the Mechanical Properties of Electrodeposits", *Plating and Surface Finishing* 77, 58 (March 1990)

19.   I. Kim and R. Weil, "The Mechanical Properties of Monocrystalline Nickel Electrodeposits", *Thin Solid Films*, 169, 35 (1989)

20.   S. Mizumoto, H. Nawafune, M. Kawasaki, A. Kinoshita and K. Araki, "Mechanical Properties of Copper Deposits from Electroless Plating Baths Containing Glycine", *Plating & Surface Finishing 73*, 48 (Dec 1986)

21.   M. Paunovic, "Significance of Tensile Testing Copper Deposits", *Plating & Surface Finishing*, 70, 16 (Nov 1983)

22.   M. Parente and R. Weil, *Plating and Surface Finishing*, 71, 114 (May 1984)

23.   I. Kim and R. Weil, "Thickness Effects on the Mechanical Properties of Electrodeposits", *Proc. SUR/FIN 88*, AESF, Orlando, Fl, 1988

24.   N.J. Petch, "The Cleavage Strength of Polycrystals", *Journal of the Iron and Steel Institute*, 174, 25 (1953)

25.   R. Walker and R.C. Benn, "Microhardness, Grain Size and Topography of Copper Electrodeposits", *Plating*, 58, 476 (1971)

26.   H. McArthur, *Corrosion Prediction and Prevention in Motor Vehicles*, Ellis Horwood Ltd., England (1988)

27.   V.M. Kozlov, E.A. Mamontov and Yu. N. Petrov., *Fiz. Metal. i Metalloved*, 26 (3), 564 (1968)

28.   G.D. Hughes, S.D. Smith, C.S. Pande, H.R. Johnson and R.W. Armstrong, "Hall-Petch Strengthening for the Microhardness of Twelve Nanometer Grain Diameter Electrodeposited Nickel", *Scripta Metallurgica*, 20, 93 (1986)

29.   A.W. Thompson and H.J. Saxton, "Structure, Strength and Fracture of Electrodeposited Nickel and Ni-Co Alloys", *Metallurgical Transactions*, 4, 1599 (1973)

30.   C.P. Brittain, R.W. Armstrong and G.C. Smith, "Hall-Petch Dependence for Ultrafine Grain Size Electrodeposited Chromium", *Scripta Metallurgica*, 19, 89 (1985)

31.   H.W. Hayden, R.C. Gibson, and J.H. Brophy, "Superplastic Metals", *Scientific American*, 220, 28 (March 1969)

32.   A. Goldberg, "Materials Engineering", *Energy and Technology Review*, Lawrence Livermore National Laboratory, 3 (March 1987)

33.   N.P. Barykin, R.Z. Valiyev, O.A. Kaybyshev and F.A. Sadykov, "Superplastic Behavior of an Electrodeposited Coating of Eutectic Alloy Cd-Zn", *Phys. Met. Metallogr. (USSR)*, 63 (2), 157 (1987)

34.   R.J. Walter, "Tensile Properties of Electrodeposited Nickel-Cobalt", *Plating & Surface Finishing*, 73, 48 (Oct 1986)

35.   R.J. Walter and H.E. Marker, "Superplastic Alloys Formed by Electrodeposition", *US Patent 4,613,388*, Sept 1986

36.   P.J. Martin and W.A. Backofen, "Superplasticity in Electroplated Composites of Lead and Tin", *Transactions of the ASM*, 60, 352, 1967

37.   M.M.I. Ahmed and T.G. Langdon, "Exceptional Ductility in the Superplastic Pb-62 Pct Sn Eutectic", *Metallurgical Transactions A*, 8A, 1832, (1977)

38.   P.S. Venkatesan and G.L. Schmehl, "Superplasticity in Metals", *The Western Electric Engineer*, Vol XV, 2 (Jan 1971)

39.   G.K. Maltseva and A.V. Belyanushkin, *Izv. Akad. Nauk SSSR*, Met. 5, 134 (1988)

40.   D. S. Lashmore, R. Oberle and M.P. Dariel, "Electrodeposition of Artificially Layered Materials", *Proc. AESF Third International Pulse Plating Symposium*, American Electroplaters & Surface Finishers Soc., (Oct 1986)

41.   D. Tench and J. White, "Enhanced Tensile Strength for Electrodeposited Nickel-Copper Multilayer Composites", *Metallurgical Transactions A*, 15A, 2039 (1984)

42.   S. Nakahara, "Direct Observations of Inclusions in Electrodeposited Films by Transmission Electron Microscopy", *J. Electrochem. Soc.*, 129, 201C (1982)

43.   J.W. Dini and H.R. Johnson, "The Influence of Nickel Sulfamate Operating Parameters on the Impurity Content and Properties of Electrodeposits", *Thin Solid Films*, 54, 183 (1978)

44.   O.B. Verbeke, J. Spinnewin and H. Strauven, "Electroformed Nickel for Thermometry and Heating", *Rev. Sci. Instrum.*, 58 (4), 654 (April 1987)

45.   J.W. Dini and H.R. Johnson, "Influence of Carbon on the Properties of Sulfamate Nickel Electrodeposits", *Surface Technology*, 4, 217 (1976)

46.   R.R. Vandervoort, E.L. Raymond, H.J. Wiesner and W. P. Frey, "Strengthening of Electrodeposited Lead and Lead Alloys, II-Mechanical Properties", *Plating 57*, 362 (1970)

47.   D.E. Sonon and G.V. Smith, "Effect of Grain Size and Temperature on the Strengthening of Nickel and a Nickel-Cobalt Alloy by Carbon", *Trans. Metallurgical Soc. AIME*, 242, 1527 (1968)

48.   J.W. Dini, H.R. Johnson and H.J. Saxton, "Influence of Sulfur Content on the Impact Strength of Electroformed Nickel", *Electrodeposition and Surface Treatment*, 2, 165 (1973/74)

49.   J.W. Dini, H.R. Johnson and H.J. Saxton, "Influence of S on the Properties of Electrodeposited Ni", *J. Vac. Sci. Technol.*, 12, No 4, 766 (July/August 1975)

50.   J.P. Chubb, "Fracture Mechanics-A Break for the Metallurgist?", *Metallurgia*, 46, 493 (August 1979)

51.   J.W. Dini, H.R. Johnson and L.A. West, "On the High Temperature Ductility Properties of Electrodeposited Sulfamate Nickel", *Plating & Surface Finishing*, 65, 36 (Feb 1978)

52.   D.H. Lassila, "Material Characteristics Related to the Fracture and Particulation of Electrodeposited Copper Shaped Charge Jets", *Lawrence Livermore National Laboratory UCRL-102520*, April 1990

53.   W.R. Wearmouth and K.C. Belt, "Electroforming With Heat Resistant Sulfur Hardened Nickel", *Plating & Surface Finishing*, 66, 53 (Oct. 1979)

54.   J.W. Dini and W.H. Gourdin, "Evaluation of Electroformed Copper for Shaped Charge Applications", *Plating & Surface Finishing*, 77, 54 (August 1990)

55.   M. Myers and E.A. Blythe, "Effects of Oxygen, Sulphur, and Porosity on Mechanical Properties of High-Purity Copper at 950C", *Metals Technology*, 8, 165 (May 1981)

56.   T.G. Nieh and W.D. Nix, "Embrittlement of Copper Due to Segregation of Oxygen to Grain Boundaries", *Met. Trans. A*, 12A, 893 (1981)

57.   L.H. Esmore, "The Electrodeposition of High Purity Chromium", *Trans. Institute of Metal Finishing*, 57, 57 (1979)

58.   D.M. Mattox, "Thin Film Metallization of Oxides In Microelectronics", *Thin Solid Films*, 18, 173 (1973)

59.   H.F. Winters and E. Kay, "Gas Incorporation Into Sputtered Films", *J. Appl. Phys.*, 38, 3928 (1967)

60.   A. Pan and J.E. Greene, "Residual Compressive Stress in Sputter Deposited TiC Films on Steel Substrates", *Thin Solid Films*, 78, 25 (1981)

61.   D.M. Mattox and G.J. Kominiak, "Incorporation of Helium in Deposited Gold Films", *J. Vac. Sci. Technol.*, 8, 194 (1971)

62.   A.J. Markworth, "Growth Kinetics of Inert-Gas Bubbles in Polycrystalline Solids", *J. Appl. Phys.*, 43, 2047 (1972)

63.   E.V. Kornelsen, "The Interaction of Injected Helium With Lattice Defects in a Tungsten Crystal", *Radiation Effects*, 13, 227 (1972)

64.   S.K. Dey and A.M. Dighe, "A Suitable Deflection Electrode Geometry for Removal of Ions Inside Deposition Chambers", *Solid State Technology*, 15, 51 (Oct 1971)

6
STRUCTURE

INTRODUCTION

Almost all plated metals are crystalline, which means that the atoms are arranged on a regular three dimensional pattern called a lattice (1). The three most important lattices are face centered cubic (fcc), body centered cubic (bcc) and hexagonal close packed (hcp), all shown in Figure 1. Face centered cubic packing of spheres often seen in fruit stands or in piles of cannonballs at war memorials, is the densest packing of spheres in three dimensional space (2,3,4). Table 1 lists the lattices for the commonly plated metals. However, it's important to note that incorporation of foreign species can modify the structure of deposited metals. For example, a structural transition from unstable hexagonal chromium hydride to body centered cubic chromium during or soon after plating accounts for the cracking observed in chromium deposits. This decomposition involves a volume shrinkage of greater than 15 percent (5). More discussion on microstructural transformations of deposits will be presented later in this chapter. Additional topics that will be covered include texture and fractals.

STRUCTURE OF ELECTRODEPOSITED AND ELECTROLESS COATINGS

The properties of all materials are determined by their structure. Even minor structural differences often have profound effects on the properties of electrodeposited metals (6). Four typical structures encountered with electrodeposited metals include; 1) columnar, 2) fibrous, 3) fine-grained, and 4) banded (7). Cross sections showing each type are

Table 1 - Lattice Structure of Commonly Plated Metals

Face centered cubic	Hexagonal close pack	Body centered cubic	Tetragonal
Ag	Cd	Cr	Sn
Al	Co	Fe	
Au	Zn		
Cu			
Ni			
Pb			
Pd			
Pt			
Rh			

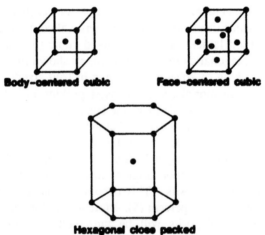

Body-centered cubic   Face-centered cubic

Hexagonal close packed

Figure 1: Unit cells of the three most important lattices.

included in Figures 2 to 6.

Columnar structures (Figure 2) are characteristic of deposits from simple ion acidic solutions containing no addition agents, e.g., copper, zinc, or tin from sulfate or fluoborate solutions, operated at elevated temperature or low current density. Deposits of this type usually exhibit lower strength and hardness than other structures but high ductility.

Figure 2: Large, Columnar Grains-cross section of a deposit produced in a citrate based acid gold solution (200 ×).

Fibrous (acicular) structures, which represent a refinement of columnar structure, are shown in Figure 3. This type of structure is obtained because some factors in the deposition process such as the presence of addition agents, or use of low temperature and high current density in copper sulfate solutions, have favored the formation of new nuclei rather than growth of existing grains. The finer grain size may be the result of interference of crystal growth by codeposited metal hydroxide or hydrogen (7). Properties of fibrous deposits are intermediate between columnar and fine-grained deposits.

Fine-grained structures (Figure 4) are usually obtained from complex ion solutions such as cyanide or with certain addition agents. These deposits are less pure, less dense and exhibit higher electrical resistivities due to the presence of codeposited foreign material. Deposits from simple ion acidic solutions, such as copper or nickel from sulfate solutions, develop this structure if operating conditions are more extreme

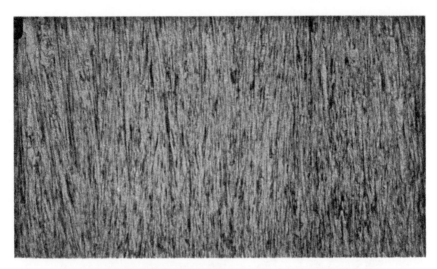

Figure 3: Fibrous (Acicular) Structure-cross section of a deposit produced in a nickel sulfamate solution (200 ×).

Figure 4: Fine Grained Structure—cross section of a deposit produced in a copper cyanide solution (200 ×).

than those that produce deposits of the type shown in Figure 3. For example, a very high current density, a high pII (in the case of a nickel solution) resulting in codeposited hydrated oxides, or certain addition agents may cause the formation of this type of structure (7). The grain sizes in deposits of this type are of the order of 10^{-5} to 10^{-6} cm. These deposits are usually relatively hard, strong and brittle but it is important to realize that some fine-grained structures can be quite ductile (30% elongation) and the grain size so small that it is virtually unresolvable, as shown in Figure 5 (8).

Laminar (or banded) structures are shown in Figure 6. The grains within the lamellae are extremely small. These structures are characteristic of bright deposits resulting from addition agents such as sulfur containing organic compounds which result in small amounts of S and C in the deposit. A number of alloy deposits such as gold-copper, cobalt-phosphorus, cobalt-tungsten, and nickel-phosphorus (electroless and electrodeposited) exhibit this structure. These deposits usually have high strength and hardness but low ductility. Similar laminations can be found in deposits produced in solutions operated with either periodic reverse current or pulse plating.

Figure 5: Very Fine Grained Structure—cross section of a deposit produced in a copper sulfate solution containing proprietary additives (500 ×).

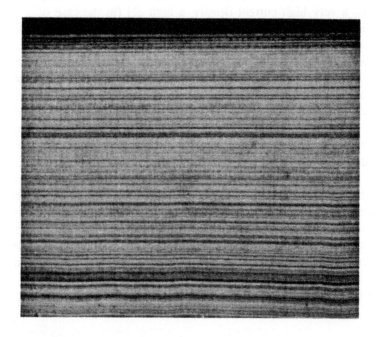

Figure 6: Laminated (or Banded) Structure—cross section of a gold-copper deposit (250 ×).

The crystal structure resulting from an electrodeposition process is strongly dependent on the relative rates of formation of crystal nuclei and the growth of existing crystals (9,10). Finer-grained deposits are the result of conditions that favor crystal nuclei formation while larger crystals are obtained in those cases that favor growth of existing crystals. Generally, a decreasing crystal size is the result of factors which increase the cathode polarization (9,10).

From the electroplaters' viewpoint, it would be nice to have some three dimensional picture that would show the influence of operating conditions on structure. However, since plating processes have numerous variables that influence structure, e.g., metal ion concentration, addition agents, current density, temperature, agitation, and polarization, a plot such as that shown in Figures 8 and 9 for physically vapor deposited films cannot be produced. However, Figure 7 does pictorially show how individual plating variables influence grain size of electrodeposits (11).

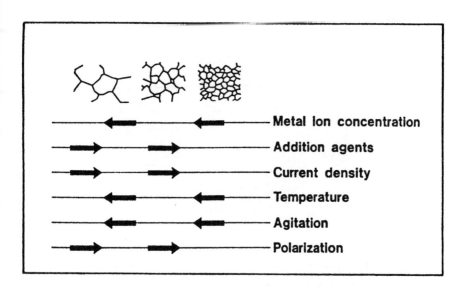

Figure 7: Relation of structure of electrodeposits to operating conditions of solutions. From reference 11.

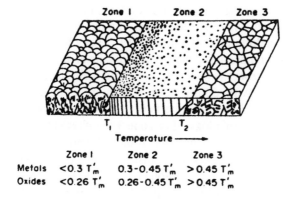

Figure 8: Structural zones in PVD films. From Movchan and Demchishin, reference 13. Reprinted with permission of Noyes Publications.

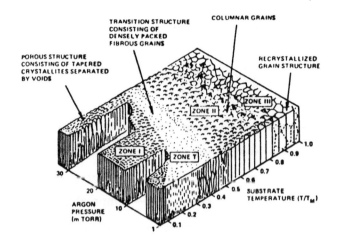

Figure 9: Structural zones in PVD films. From Thornton, references 14, 15. Reprinted with permission of Noyes Publications.

STRUCTURE OF PHYSICALLY VAPOR DEPOSITED COATINGS

With physically vapor deposited (PVD) coatings there have been three distinct steps taken in the classification of thin film morphology (12). Movchan and Demchishin (13) were the first to classify thin films using a structure zone model (SZM). They observed that regardless of the thin film material, its morphological structure is related to a normalized, or reduced, temperature T/Tm, where T is the actual film temperature during deposition, and Tm is its melting point, both in K (12). They found that by increasing the deposition temperature, they could obtain at least three qualitatively distinct structure zones (Figure 8). Zone 1 in their classification consists of tapered columns with domed tops and is in a region of low adatom mobility. In zone 2, the structure is of a straight columnar nature and has a smooth surface morphology. For zone 3 the physical structure resembles equiaxed crystallites, much the same as those found in recrystallized metals.

Unlike Movchan and Demchishin (13) who prepared their films by electron beam evaporation, Thornton (14,15) used magnetron sputtering and introduced a new parameter, the sputtering gas pressure. He showed that both T/Tm and the sputtering gas pressure have an identifiable and significant effect on thin film growth (12). Thornton's model includes a fourth transition zone, called zone T, between zones 1 and 2 (Figure 9). In this zone the films have a smoother surface morphology and are denser than films from the surrounding zones (12).

Recently, Messier and colleagues, have shown that the physical structure of thin films changes as a function of thickness (12,16-19). A distribution of sizes from the smallest clustered units (µm-sized) to the largest, dominant sizes perceived, typically µm-size units in SEM micrographs, is the resulting structural heterogeneity (16). Thornton's model is essentially retained in this new SZM which includes the similarity in morphology of various levels of magnification as well as the evolutionary growth of morphology (17). A revised SZM model for zone 1 structures is shown in Figure 10, wherein all the distinct levels of physical structure column/void sizes are considered and assigned subzones 1A, 1B, 1C, 1D and 1E (17). The smallest size level (1-3 µm) is represented by zone 1A and the largest by zone 1E (300 µm column sizes). Larger sizes can be assigned designations of 1F, 1G, etc. This structure is not unique to the deposition technique but has been found in all vapor deposited films, as well as electrodeposited films (12,18). This universality in the physical structure of a variety of materials and self-similarity in structural evolution indicates a common origin of thin film growth and a possible fractal description (12,16,18,19). Fractals are discussed later in this chapter.

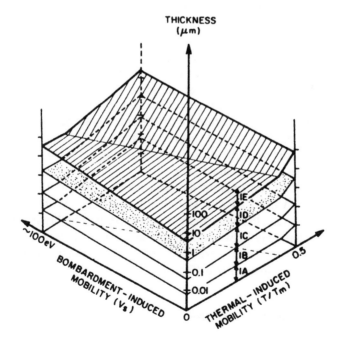

Figure 10: Revised structure zone model for films. From Messier, Giri and Roy, reference 17. Reprinted with permission of the American Vacuum Society.

INFLUENCE OF SUBSTRATE

The structure of most electrodeposits is determined by epitaxial and pseudomorphic growth onto a substrate and by the conditions prevailing during deposition. Typically, a deposited metal will try to copy the structure of the substrate and this involves epitaxy, which occurs when definite crystal planes and directions are parallel in the deposit and substrate, respectively (1,20). Epitaxy is the orderly relation between the atomic lattices of substrate and deposit at the interface, and is possible if the atomic arrangement in a certain crystal direction of the deposit matches that in the substrate. Another term, pseudomorphism, refers to the continuing of grain boundaries and microgeometrical features of the cathode substrate into the overlying deposit. A deposit stressed to fit on the substrate is said to be pseudomorphic (20). Pseudomorphism persists longer than epitaxy.

The structure of a deposit and its properties and adhesion can be noticeably influenced by the substrate upon which it is plated. Figure 11 shows cross sections of copper electrodeposited on cast copper (9). If the substrate was cleaned but not pickled prior to plating, the structure of the plated deposit was quite different (fibrous) compared to that of the cast copper (coarse grained), Figure 11a. However, use of pickling after cleaning resulted in a structure wherein the copper crystals were continuations of the crystals in the copper basis metal (Figure 11b). Such reproductions of the basis metal structure may occur even with dissimilar metals that may vary appreciably in lattice structure and spacing (21).

The effect of the type of substrate on the properties of nickel electrodeposited on as-rolled and on annealed, cube-textured copper sheet is shown in Table 2. The influence of the small grain size induced in the deposits plated on the as-rolled sheet is apparent in the higher strength and ductility, compared with the deposit plated on the annealed, cube textured sheet which was coarse-grained (22)

Figures 12 and 13 show the influence of substrate on elongation of copper deposited from an acid sulfate solution (23). With 304 stainless steel as the substrate, the elongation of the copper was highly irreproducible and drifted alternately between high and low values (Figure 12). Acceptably reproducible results were obtained with a much more corrosion resistant substrate, Inconel 600 (Figure 13).

PHASE TRANSFORMATIONS

A phase transformation is a change in the number or nature of phases as a result of some variation in the externally imposed constraints such as the temperature, pressure, or magnetic field. As will be shown in

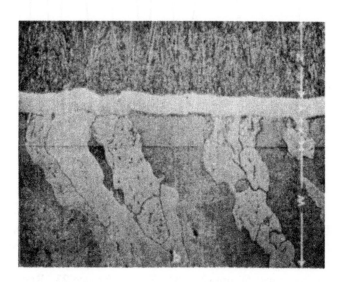

Figure 11: Cross section of the chilled side of cast copper, upon which copper was electrodeposited. (a) The surface was cleaned, but not pickled prior to electroplating. (b) The surface was pickled (bright-dipped) after cleaning. The various zones in (b) also apply to (a), i.e., W, the base metal; X, the electrodeposited copper (first layer); Y, electrodeposited nickel; and Z, electrodeposited copper (second layer). Reprinted with permission of McGraw-Hill, Inc.

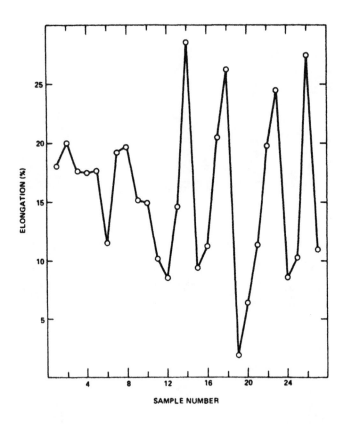

Figure 12: Measured elongation for a series of deposits plated on a rotating (750 rpm) 304 stainless steel mandrel from an acid copper sulfate solution containing 5 ml/l CUBATH M HY additive (Oxy-Metal Industries). Deposit thickness was 50 um (2 mils). From reference 23. Reprinted with permission of American Electroplaters & Surface Finishers Soc.

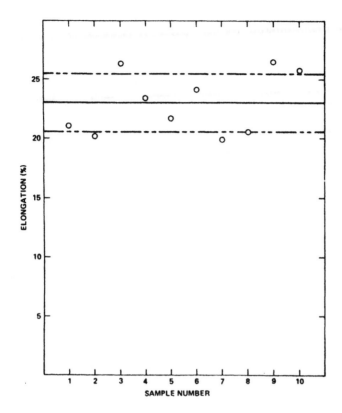

Figure 13: Elongation for a series of deposits plated on an Inconel 600 mandrel under the same conditions as Figure 12.   From reference 23. Reprinted with permission of American Electroplaters & Surface Finishers Soc.

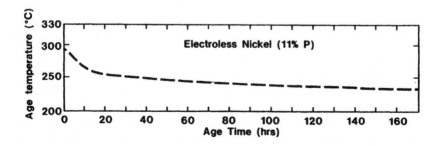

Figure 14: Time/temperature profile illustrating the transition from an amorphous to a crystalline structure for electroless nickel containing 11% (wgt) phosphorus. Adapted from reference 29.

Table 2 - Effect of Substrate on Properties of Electroplated Nickel* (from Reference 22)

Material	Composition	Young's Modulus GPa	Yield Strength MPa	Tensile Strength MPa	Percent Elongation
Ni plated on as-rolled Cu	Ni	104	215	382	2.2
		5**	6	34	0.3
Ni plated on cube-textured Cu	Ni	104	120	222	7.1
		4**	3	7	0.5

*   Nickel was plated in an all-sulfate solution at a current density of 2.5 mA/cm^2 to a thickness of about 10 µm.

** Values in this line are standard deviations.

the following examples, such changes in the nature of either a substrate or a coating after being deposited on a substrate can noticeably affect properties.

A. Electroless Nickel

As-deposited electroless nickel is metastable and undergoes a crystalline transition at moderate temperature (240 to 400°C). This change causes a rapid increase in the hardness and wear resistance of the coating while reducing the corrosion resistance and ductility (24,25). The transition also causes an increase in density and accordingly a decrease in volume. This volume change, which can vary from 0.1% to 1.3% (26-28), coupled with differential thermal expansion, is the cause for cracking or fissuring often found in deposits after heat treatment (28).   The extent of the crystalline transition is a complex function of a number of factors including: 1) temperature and time at temperature, 2) heating rate, 3) previous temperature history, and 4) phosphorus content (29). Figure 14 provides a time/temperature profile which illustrates the transition from an amorphous to a crystalline structure for electroless nickel containing 11% phosphorus. If thermal exposure is maintained in the time/temperature envelope below the dotted curve, then the electroless nickel will remain entirely amorphous. However, if exposure conditions fall above this curve, then partial or complete crystallization will occur (29).

B. Gold-Copper

The gold-copper system exhibits disorder-order transformation and the strengthening mechanism of these electrodeposited alloys is associated with this phenomenon. A 50 at% Au-Cu (75 wt% Au) alloy has a face centered cubic structure in the disordered state. In the ordered condition it has a tetragonal structure. During transformation there is a volume decrease which results in the formation of lattice strains. Increases in mechanical strength during heat treatment correlate with increases in coherency strains between ordered nuclei and disordered matrices (30). Longer annealing times eliminate these strains by microtwinning and result in decreases in mechanical strength. Typical results obtained with gold-copper electrodeposited alloys are shown in Figure 15.

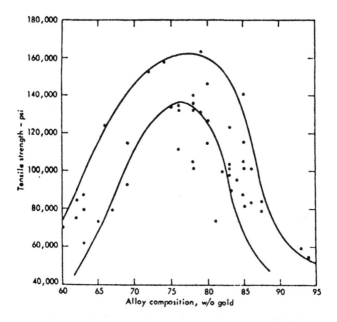

Figure 15: Effect of heating for 3 hours at 350°C on the tensile strength of electroformed gold-copper alloy. From reference 30. Reprinted with permission of American Electroplaters & Surface Finishers Soc.

C. Transformation in Tin-Nickel

Electrodeposited tin-nickel (NiSn) is an another example of a deposit that undergoes a transformation upon heating.This phase change causes cracking and exfoliation of the coating, so that while actual melting of the deposit does not take place below about 800°C, the electrodeposited material cannot be safely used at temperatures above about 250°C (31). NiSn transforms to Ni_3Sn_2 and Ni_3Sn_4 when isothermally heated at 300°C for at least 20 hours and when isothermally heated at and above 400°C for at least 1 hour. The deposit has been shown to be stable at room temperature for at least 10 years (32). It is not recommended as a diffusion barrier for parts subjected to high temperatures. For example, tin-nickel deposits 12.5 and 2.5 μm thick, deposited as a diffusion barrier between 60 μm fine gold and copper showed decomposition into discrete particles after only 50 hours exposure at 500°C. Gold and copper completely penetrated the original tin-nickel layer as if it were not even present(33).

D.   Palladium

An unstable thermodynamic phase resulting from high hydrogen in the lattice can cause cracking of palladium deposits during or after plating thus exposing the less noble substrate and significantly reducing corrosion performance (34,35). The key is keeping the H/Pd ratio below 0.03 since in this range palladium hydride is present in an α phase with a lattice constant close to that of pure Pd. Escape of hydrogen from this structure does not cause any lattice distortion. When the H/Pd ratio is above 0.57 palladium hydride is present in a β phase with a lattice constant which is about 3.8% higher than that of the α phase. The β phase is thermodynamically unstable and converts to α phase with the release of hydrogen and causes a contraction of the lattice and cracking of the deposit. Deposits with a H/Pd ratio between 0.03 and 0.57 have a combination of α and β phases and since the β phase causes problems, the way to avoid subsequent cracking of deposits is to keep the H/Pd ratio below 0.03.

E. Cobalt

Cobalt exhibits an unusual structural transformation (fcc → hcp) as a result of hydrogen inclusion and subsequent outdiffusion. A peculiar feature of cobalt deposits is that a high temperature phase (>417°C) can be obtained by deposition at ambient temperature. This "fcc cobalt" is produced in a solution with a pH less than 2.4 by the simultaneous incorporation of hydrogen to form an "fcc cobalt hydride". Since this hydride is unstable at ambient temperatures, it decomposes into basic cobalt (hcp) during the outdiffusion of hydrogen. At high pH (>2.9), stable hcp cobalt is obtained (36). Incorporation of hydrogen and hydroxide also significantly alters the microstructure of nickel deposits (37).

F. Miscellaneous

The bores of many cannon tubes are electroplated with chromium to provide better resistance to erosion, wear and corrosion resistance (38). In fact, the resistance of the chromium deposit to erosion and wear is so good that there is little or no wear of the bore until the chromium begins to spall off the steel substrate. This spalling appears to chiefly be the result of the underlying steel undergoing a phase transformation with an attendant volume increase. It has been found that if the electroplate is made thicker, the underlying steel never reaches the transformation temperature and the useful life of the plated cannon is considerably increased (38).

Metal surfaces exposed to gas may undergo transformation, with deterioration of their properties. The rate of the phenomenon depends largely on the nature of the metals, the attacking gases, and the new products that may form at the interface between the two phases (39). One example is silicon which upon oxidizing to SiO_2 undergoes expansion resulting in highly compressive stresses on its surface (40).

MICROSTRUCTURAL INSTABILITY AT ROOM TEMPERATURE

A. Copper

The metallurgical instability of some copper deposits can create problems when this material is used for optical applications. Copper deposits have been shown to markedly soften after storage at room temperature for 30 days (41). Another example relates to copper mirrors which revealed a change in their optical surfaces over a period of six months. This was caused by recrystallization which had occurred in the copper attendant with a shifting of the surface along individual grain boundaries (42). More information on this is presented later in this chapter in the discussion on texture. After recrystallization, an orange-peel effect is visible to the eye in some cases. Figure 16 shows a diamond turned copper surface before recrystallization and Figure 17 shows a surface after recrystallization. The problem is caused by the high density of defects in the electroplated copper, often much higher than that achieved by cold working (discussed in more detail in the chapter on Properties). With copper, the problem is exacerbated by lower current densities (less than 20 asf).

Figure 16: Normarski micrograph of a diamond turned copper surface before recrystallization (100 ×). From reference 42. Reprinted with permission of the American Electroplaters & Surface Finishers Soc.

Figure 17: Normarski micrograph of a diamond turned copper surface after recrystallization (100 ×). From reference 42. Reprinted with permission of the American Electroplaters & Surface Finishers Soc.

Differential scanning calorimetry (DSC)[1] reveals that copper plated at 5 asf has a very high recovery energy at 148°C and a much lower recrystallization energy exotherm at 284°C (Figure 18, Ref. 42). By comparison, copper plated at 15 asf shows a joined recovery/ recrystallization energy exotherm at 306°C (Figure 19). The way to eliminate the recrystallization problem at room temperature is to heat treat at a low temperature (1 hour at 250°C). This removes the recovery energy and no recrystallization and grain growth occurs (42). Figure 20, a DSC of copper plated at low current density (5 asf) and then heated for 1 hour at 250°C, shows no retained recovery energy and a reduction in the recrystallization energy.

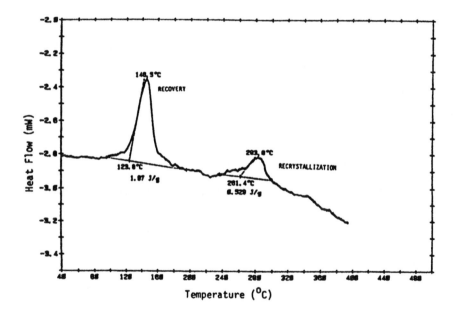

Figure 18: Differential scanning calorimetry plot showing two exotherms for copper plated at 5 asf. From reference 42. Reprinted with permission of the American Electroplaters & Surface Finishers Soc.

---

[1]   Differential scanning calorimetry (DSC) measures the heat absorbed or released by a substrace as it passes through transitions or undergoes reactions. DSC can pinpoint the exact temperature or time in a process when a material goes through a transition, when it occurs,and where decomposition occurs.

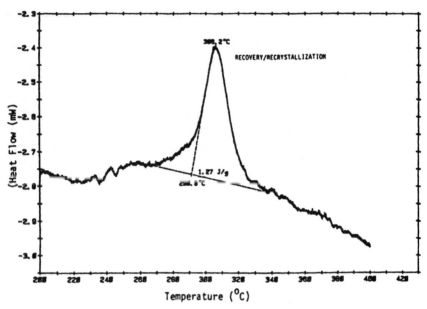

Figure 19: Differential scanning calorimetry plot showing energy release for copper plated at 15 asf. From reference 42. Reprinted with permission of the American Electroplaters & Surface Finishers Soc.

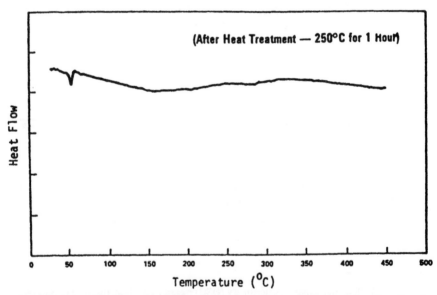

Figure 20: Differential scanning calorimetry plot of copper plated at 5 asf and then heat treated. From reference 42. Reprinted with permission of the American Electroplaters & Surface Finishers Soc.

B. Silver

A silver flat mirror turned to better than 0.5 fringe deformed 9 fringes in 15 months after machining. This was due to room temperature creep (relief) of the silver electroplating stress. Other examples are brass mirrors which had been diamond turned flat, checked in an interferometer and electroplated with silver. The plating operation caused them to deform by 10 fringes. Removing the silver by machining eliminated the distortion and essentially restored the original features. The problem was eliminated by heat treating copper or silver plated parts at 150°C for 1 hour prior to diamond turning (43). Table 3 shows that some silver deposits exhibited a significant decrease in hardness as a function of time at ambient temperature after plating. This decrease in hardness has been attributed to recrystallization and grain growth(44).

Table 3 - Hardness of Electrodeposited Silver from Cyanide Electrolytes as Related to Storage Time Versus Solution Additive[a]

Solution Additive	Deposit Hardness (VHN)				
	Initial	1/2 Mo.	1 Mo.	3 Mo.	6 Mo.
None	110	105	102	100	90
Carbon Disulfide, 1-1/2 ml/l	148[b]	--	135	130[b]	130
Sodium Selenate, 0.8 g/l	125	124	107	65	66
Potassium Sulfate, 50 g/l	140	132	105	95	85
Cobalt, 1.7 g/l	108	98	87	84	88

(a) From reference 44; the solutions contained 26 to 30 g/l silver, 20 to 42 g/l free potassium cyanide and 30 to 34 g/l potassium carbonate. Plating current density was 21.5 A/m².

(b) Grain size increased from 2.36 to 3.22 x 10^{-6}cm during three months storage.

TEXTURE

Texture, which is preferred distribution of grains (individual crystallites) having a particular crystallographic orientation with respect to a fixed reference frame, is an important structural parameter for bulk materials and coatings (45,46). It can noticeably affect a variety of functional properties which will subsequently be discussed and is an excellent materials science example illustrating how properties can be tailored. However, before discussing texture in detail, a brief review on crystallography is in order. It is convenient to be able to specify certain planes in crystal lattices and there are sets of conventional descriptors for this (4,20,47). Parentheses (100) around Miller indices signify a single plane or set of parallel planes. Curly brackets or braces {100} signify planes of a "form" - those which are equivalent in the crystal - such as the cube faces of a cubic crystal. To distinguish the notation for a direction from that of a position, the direction integers are enclosed in square brackets [100]. A set of directions which are structurally equivalent and called a family of directions is designated by angular brackets <100> (4,47). The three most important planes are {100}, {110} and {111} shown for cubic lattices by the hatched lines in Figure 21, which also includes the three most important directions, namely, <100>, <110>, and <111> (20).

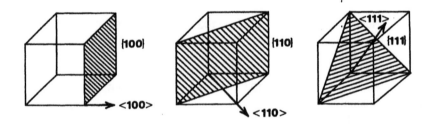

Figure 21: Crystal planes and directions. (a) {100} planes are faces of the cube, <100> directions are cube edges. (b) The sides of {110} planes are two face diagonals and two cube edges, <110> directions are face diagonals. (c) The sides of {111} planes are face diagonals, <111> directions are cube diagonals. One plane of each type is shown hatched. One direction of each type is shown by an arrow. Adapted from reference 20.

Table 4, which lists textures of coatings produced by various techniques, reveals that texture is highly dependent on the coating process used. For that matter, textures of electroplated coatings can be markedly influenced by solution composition and operating conditions. Examples include Pd, Ni and Au. With Pd, the texture of coatings for an identical coating composition can be totally different and vary from textureless to a pronounced (111) or (110) texture or a combination of (111), (100), (211) and (110) components (45). Depending on pH and current density, five different textures can be obtained from a Watts nickel plating solution as shown in Figure 22 (48). Gold coatings produced on a rotating disc electrode in a cyanide solution show a variety of textures as an influence of current density and deposition rate (Figure 23a). Codeposition of particulates with the gold deposit shifts the texture domains (Figure 23b) thus indicating that composite layers have their own structural properties (45).

Table 4 - Texture of some coatings produced by various techniques*

Process Material	Electro-deposition (25 - 100°C)	PVD (200-400°C)	CVD (>500°C)
Ag	(111),(100)	(111)	
Cr	(111),(100)	(110)	
Fe	(111),(100)	(211)	
Au	(111),(211)	(111)	
NiFe	(111),(100)	(111)	
TiN		(111),(200)	
TiC		(111),(100)	(110)(111)(200)

* From reference 45

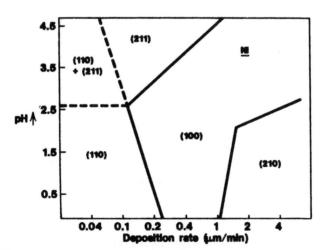

Figure 22:   Texture stability diagram for nickel deposited in a Watts solution. Adapted from reference 48.

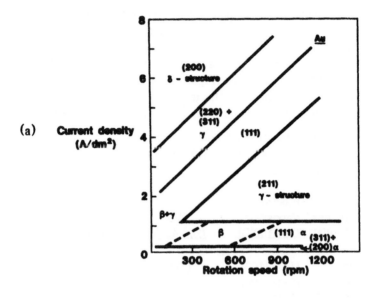

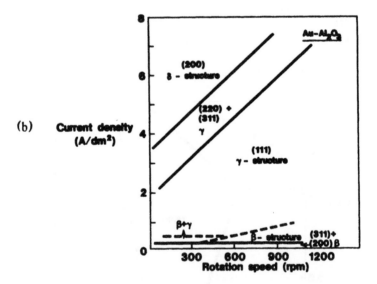

Figure 23: Texture stability diagrams for (a) pure gold coatings produced in a cyanide solution and (b) composite gold coatings.   Adapted from reference 45.

Texture is completely independent of the substrate orientation for thick deposits (several µm or 10 µm, depending on plating conditions) (48). Electrochemical parameters appear to be the only governing factor. For example, texture mainly depends on cathodic potential and pH of the solution for a given electrolyte composition. This also applies to current density if temperature is kept constant (48). Some information on substrate texture and electroplated metal textures is shown in Table 5. This reveals that electrodeposits have the <111> direction normal to the surface for BCC crystal structures and the <110> direction for FCC substrates, independent of substrate orientation. With hexagonal closed packed metals such as zinc, the <10$\bar{1}$2> direction is predominant and with tetragonal metals such as tin, the <231> direction is preferred (49).

In the case of physical vapor deposition, the deposition rate directly determines the texture of coatings (45). For example, with HfN coatings, a distinct change from a predominant (111) to a (311) texture is noticed with increasing deposition rates (49,50).

Table 5 -   Relationship between the Substrate and the Electroplated Metal Textures[1]

Crystal Structure	Substrate Texture	Electroplated Metal Texture
	Cu <110>	Fe <111>
BCC	Cu <111> + <110>	<111>
	Au <111>	<111>
	Ni <110>	Cu <110>
FCC	Au <111>	<110>
	Fe <111>	<110>
Hexagonal	Fe <100>	Zn <10$\bar{1}$2>
Close Packed	<110>	<10$\bar{1}$3>
Tetragonal	Cu <111> + <100>	ß-Sn <231>
	Au <111>	<231>

(1) From Reference 49.

INFLUENCE OF TEXTURE ON PROPERTIES

Texture can noticeably affect a variety of properties such as formability, corrosion resistance, etching characteristics of copper foil, paint adhesion on zinc, contact resistance, magnetic properties, wear resistance, porosity, and hardness of copper deposits. Examples will be discussed in the following sections.

A. Formability

Electrogalvanized (zinc plated) steel is used by the automotive industry to improve the corrosion resistance of body panels. The sheet is exposed to a variety of processes during manufacturing. These include blanking, forming, spot welding, phosphating, and painting. A deposit structure compatible with these processes is crucial to the success of industrial electrogalvanizing. By adjusting the plating parameters to impart a desirable texture to the zinc layer, formability of electrogalvanized steel can be improved and cracking of the deposit prevented.

Plastic deformation in crystalline solids involves slip which is the displacement of one atom over another by dislocation movement. The major slip plane for zinc, which has a HCP lattice, is the basal plane (0001). In this plane, slip is at least 30 times easier than for any other in the HCP system (52,53). Basal plane deposits which are plastically deformed would exhibit elongation of individual grains, causing one layer of grains to slide over the next as shown in Figure 24a. Continuity of the coating is preserved by this action and this is essential for corrosion resistance. With prism-plane (10$\bar{1}$0) deposits, fracture would likely occur since the stress is normal to the slip plane (Figure 24b). Results of drawbead tests show the fracture that is dominant in pyramid plane deposits which exhibit a pattern of parallel cracks normal to the draw direction. As shown in Figure 25, the deposit with the basal planes aligned parallel with the substrate exhibited good ductile elongation while the deposit with the basal planes nearly normal to the steel cracked under the same deformation (54,55). In practice the results aren't quite so simple since the stamping section is deformed by convex bending, concave bending and stretching (54). For more detail on texture and formability of zinc coated steel see references 46 and 56.

Plating experiments have revealed that pH strongly influences orientation of zinc deposits while temperature has a lesser effect (Table 6). The influence of flow rate is negligible (54).

B. Corrosion

A metal surface is a complex of crystal faces, edges, corners,

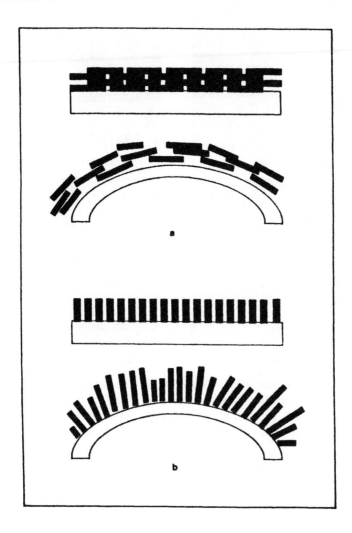

Figure 24: Consequences of simple tensile bending on deposits of (a) complete basal (0001) and (b) prism (10$\bar{1}$0) plane orientation. From reference 54. Reprinted with permission of the American Electroplaters & Surface Finishers Soc.

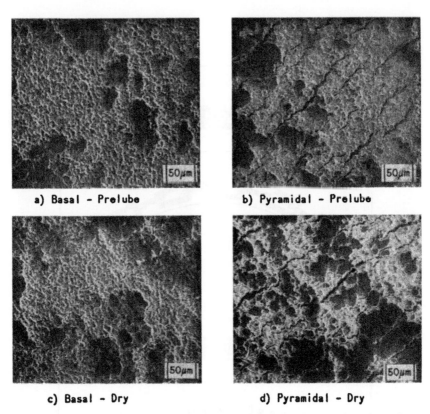

a) Basal - Prelube   b) Pyramidal - Prelube

c) Basal - Dry   d) Pyramidal - Dry

Figure 25: SEM photographs showing orientation groups of electrogalvanized zinc surfaces after drawbed tests: (a) basal, prelube; (b) pyramidal, prelube; (c) basal, dry; (d) pyramidal, dry. From reference 54. Reprinted with permission of the American Electroplaters & Surface Finishers Soc.

Table 6 - Zinc Orientation Under Selected Deposition Conditions*

Current density A/dm^2	Temp C	pH	Basal (%) (0001)	Pyramidal(%) (10$\bar{1}$X)
150	50	3.0	80-95	5-20
150	50	4.2	>45	>45
50	26	3.5	>60	32
50	26	4.5	20	>75

* From Reference 54

boundaries and disturbed layers (57). A composite of these various types of structure, in turn, dictates the properties of the surface. The change in free energy determines the reaction rate and this varies with the crystal face because the arrangement of atoms is different with different crystal faces. For example, with tungsten the ratio of free energies between two faces may in some cases be as high as a factor of 5. A practical example of this is shown in Table 7 which lists the relative rates of etching on various faces of copper crystals by acids containing hydrogen peroxide. With the rate of attack on the (111) face taken as unity, it can be seen that the etching rate can vary by as much as a factor of almost 2 to 1 (57).

Table 7.   Relative Rates of Corrosion of Different Faces of Copper Crystals*

	Crystal Face		
	(111)	(100)	(110)
0.3 N HCl – 0.1 N H$_2$O$_2$	1	0.90	1.0
0.3 N Acetic acid -- 0.1 N H$_2$O$_2$	1	0.90	0.55
0l.3 N Propionic acid -- 0.1 N H$_2$O$_2$	1	1.28	1.33

* From reference 57

With nickel, the rate of anodic solution of different grains increases in the order (111) <(100) <(110) (58). With an increase in the degree of cold deformation there is an increase in the anodic current density in the active area (59).

The corrosion behavior of zinc alloy coatings is affected by texture in terms of the ability of the layer to act either sacrificially or as a protective corrosion barrier. For example, zinc grains with a near {0001} orientation have much lower corrosion currents in sodium hydroxide solutions than those of other orientations (60). Corrosion rate data in 0.5 N sodium hydroxide solution for polycrystal and oriented crystal specimens are shown in Table 8. The corrosion current rate decreased as the packing density increased (61). Others have also discussed the influence of texture on the corrosion of zinc deposits (54,62).

Table 8 - Corrosion Potentials, Currents and Rates for Zinc*

Exposed Crystal Plane	Corrosion Current μ amps/cm^2	Calculated Corrosion Rate mil/year
(0001)	81	48
(10$\bar{1}$0)	127	75
(11$\bar{2}$0)	261	153
Poly-crystal	160	94

* From Reference 61

C. Etching Characteristics of Copper Foil

The simple etch process that has worked so well for years in printed wiring board fabrication is becoming less viable as geometries decrease due to the inherent undercut associated with the process.   Isotropic etch processes deliver copper conductors having approximately trapezoidal cross sections and irregular widths (63).   An anisotropic etch chemistry has been developed for producing straight sidewall etched structures with little or no line loss from copper foils which have a high (111) component. Figure 26 illustrates the relationship between line loss upon etching (undercut) and (111) content of the foil substrate (64). However, it's important to note that experiments with other types of copper foils showed that the correlation given in Figure 26 is only valid for copper foils electrodeposited on aluminum carriers.   Copper foils prepared by other methods etched isotropically regardless of (111) content.   X-ray pole figure analysis revealed significant qualitative structural differences between the anisotropically etching foils and the other high (111) foils which etched isotropically (63).

Others have reported that copper crystal orientation is only a factor in the definition of fine line features (3 mils or less) and that the best and worst foils relative to (111) content differed by at most 0.5 mils. This difference, in their observation, is negligible or becomes incorporated in other process variations when dealing with line widths of 4-5 mils, or greater (65).

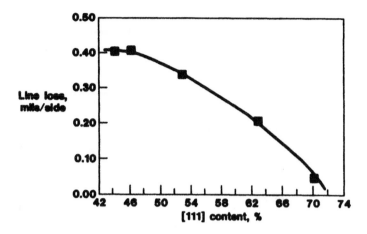

Figure 26: Relationship between line loss upon etching (undercut) and (111) content of copper foil. Adapted from reference 64.

D. Paint Adhesion on Zinc Coated Steel

Adhesion of paint to zinc coated steel is greatly affected by texture (66,67). Zinc grains with {0002} orientations parallel to the sheet surface provided better paint adherence than those of other orientations (Figure 27). This was related to two factors: 1) on deformation these grains would not yield in tension due to bending because the slip plane {0001} was parallel to the surface (the bending tension causes many small microcracks to occur in the zinc layer thus relieving residual stresses between the paint and zinc grains), and 2) zinc with a large fraction of {0001} oriented grains was more readily cleaned of contaminating organics that might have conceivably interfered with paint adherence (49,66,67).

E. Contact Resistance of Electroplated Nickel

Various nickel platings have been shown to provide surprisingly different contact resistances when exposed to the atmosphere for long periods of time (68). The only difference that could be observed among the various deposits was a variation in plating texture. Force vs. contact resistance curves for various orientations of single crystal nickel are shown in Figure 28 and verify that the difference in contact resistance can be explained by the orientation and its effect on oxide growth. Polycrystalline nickel electrodeposits were shown to behave similarly to single crystals, as

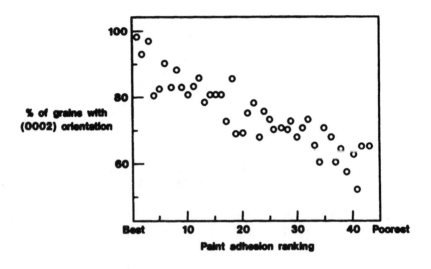

Figure 27: Relationship between the orientation of zinc and paint adhesion. Adapted from reference 67.

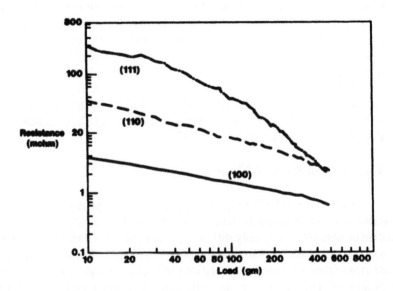

Figure 28: Resistance-force curves for three oriented nickel single crystal surfaces. Adapted from reference 68.

long as a majority of the grains were oriented with the {100} plane parallel to the surface. The oxidation rate of {100} oriented nickel single crystal surfaces was shown to be self limiting at room temperature. Figure 29, a probability plot of a heat age test, provides information on contact resistance. Change in resistance data are plotted as a cumulative percentage for three electrodeposits: nickel with a {100} texture, nickel with a {111} texture and gold. Ideal behavior would be represented by a horizontal line while deviations from this, particularly a significant upward trend, would indicate degradation.   Figure 29 shows that nickel with a {100} texture behaves slightly better than gold plated samples and significantly better than {111} textured nickel (68).

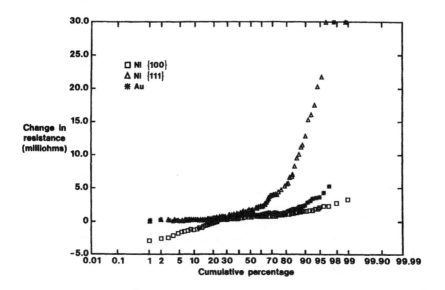

Figure 29:   Change in resistance vs. cumulative percentage for {100} nickel, {111}, nickel, and gold after 103 days heat age testing.   Adapted from reference 68.

F. Magnetic Properties of Cobalt/Phosphorus Films

Electroless plated cobalt/phosphorus thin films are usually characterized by low to medium coercivity values (300-900 Oe). For some applications, it would be desirable to have films with coercivities greater than 900 Oe. Table 9 shows that films with minimum preferred orientation exhibit maximum coercivity values (69).The ratio of the intensity of the (002) and (101) diffraction rings changes with the hypophosphite concentration of the plating solution with zero preferred orientation obtained at solution hypo concentrations of 5-6 v/o.

Table 9 - Influence of Solution Hypophosphite Concentration on Preferred Orientation and Magnetic Coercivity of Electroless Cobalt/ Phosphorus Films*

Solution hypophosphite concentration (v/o)	Preferred orientation (002)/(101)(%)	Hc (Oe)
5	0	1026
6	0	938
4	<10	930
3	<10	745
7	<10	461
2	>50	493
8	>50	200

*From reference 69.

G. Wear Resistance

With face centered cubic crystals the slip plane, which is the crystallographic plane in which adjacent atomic layers can slide over each other with minimum friction, is the (111). A face centered cubic electrodeposit such as palladium is typically (110) rather than (111) oriented and does indeed exhibit a large amount of wear when tested against itself in sliding friction.   A (111) oriented deposit, produced by revising the brightener system exhibits superior wear performance (70).   The wear resistance of ion plated TiN or TiC films is also related to preferred orientation, increasing as the intensity diffracted from the (111) plane becomes stronger (71,72).

H. Porosity

Rates of pore closure and covering power for bright gold deposits on copper are related to the crystallographic orientation of the deposit. This is discussed in detail in the chapter on Porosity and also illustrated in Figures 13 and 14 in that chapter. Similar effects have been shown with palladium electrodeposits as indicated in Figure 30.   The curve for the higher degree of preferred orientation (111), which for face-centered cubic palladium is the most densely packed, is displaced downwards indicating reduced porosity when compared with the deposit that is (110) oriented (70).

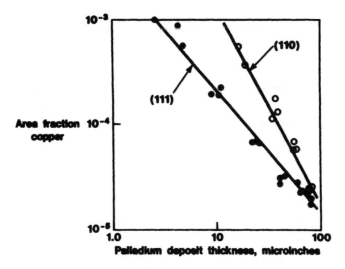

Figure 30: Porosity versus thickness plots for palladium electrodeposits from chelated solution on copper electrical terminals. Porosity measurements by corrosion potential in 0.1 M ammonium chloride. Adapted from reference 70.

I. Electrodeposited Copper Recrystallization

The metallurgical instability of some copper deposits at room temperature was mentioned earlier in this chapter. As discussed in the chapter on Additives, organic addition agents are often used to help produce bright copper deposits. Components or break-down products from these additives can be incorporated in the deposit and provide specific growth orientation as well as stresses. In an effort to form a more thermodynamically stable structure, recrystallization is likely to occur (45,73). This recrystallization is accompanied by a texture evolution and well as a lowering of hardness as shown in Figure 31. The relationship between hardness and [111] content of a copper foil is shown in Figure 32.

FRACTALS

A. Introduction

Deposits produced in solutions containing no additives, particularly simple salt solutions, are often dendritic or treelike and aside from artistic uses, of no practical value until recently. Now these dendritic structures are playing an important role in the new science of fractals (74). Figure 33a shows the structure obtained in a zinc sulfate solution containing no

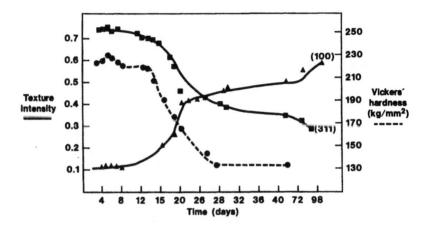

Figure 31: Change with time, at room temperature, of the intensity of texture components in copper electrodeposits. Adapted from reference 73.

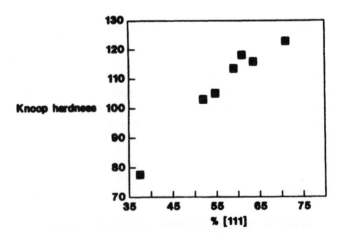

Figure 32: Relationship between Knoop hardness and the [111] content of copper foil. Adapted from reference 64.

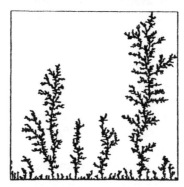

Figure 33: (a) Electrodeposited clusters (about 5 mm long), photographed 15 min. after the beginning of the growth. (b) A diffusion-limited aggregate computed with a random walker model. The digitized images in both (a) and (b) have about 1.6×10^4 boundary sites. From reference 74. Reprinted with permission of the American Physical Society.

additives, and Figure 33b a fractal"tree" grown by a computer algorithm called diffusion-limited aggregation. An interesting observation is that Figure 33b bears a striking resemblance to "trees" or "dendrites" produced by electrolysis in simple salt solutions (74).

A fractal is an object with a sprawling tenuous pattern (75). Magnification of the pattern would reveal repetitive levels of detail; a similar structure would exist on all scales. For example, a fractal might look the same whether viewed on the scale of a meter, a millimeter or a micrometer. Examples of fractals in nature include, formation of mountain ranges, ferns,coastlines, trees, branching patterns of rivers and turbulent flow of fluids or air (75,76). In the human body, fractal like structures abound in networks of blood vessels, nerves and ducts. Airways of the lung shaped by evolution and embryonic development, resemble fractals generated by a computer (77).

Although the entire field of fractals is still in its infancy, in many instances applying either theoretical fractal modeling and simulations or performing fractal analysis on experimental data, has provided new insight on the relation between geometry and activity, by virtue of the very ability to quantitatively link the two (78). For the physiologist, fractal geometry can be used to explain anomalies in the blood flow patterns to a healthy heart. Studies of fractal and chaos in physiology are predicted to provide more sensitive ways to characterize dysfunction resulting from aging, disease and drug toxicity (77). For the materials scientist, the positive aspect of fractals is that a new way has been found for quantitative analysis of many microstructures of metals. Prior to this only a quantitative description has been available. This offers the potential for a better understanding the origin of microstructures and the bulk properties of metallic materials (79).

Fractal geometry forms an attractive adjunct to Euclidean geometry in the modeling of engineering surfaces and offers help in attacking problems in tribology and boundary lubrication (80). Fractured surfaces of metals can be analyzed via fractal concepts (77,81-88). Interestingly, the term "fractal" was chosen in explicit cognizance of the fact that the irregularities found in fractal sets are often strikingly reminiscent of fracture surfaces in metals (81).

For the coater, besides the items mentioned above, fractal analysis provides another tool for studying surfaces and corrosion processes (89-91). As Mandelbrot, the father of fractal science, wrote, "Scientists will be surprised and delighted to find that not a few shapes they had to call grainy, hydralike, in between, pimply, ramified, seaweedy, strange, tangled, tortuous, wiggly, wispy, wrinkled and the like, can, henceforth, be approached in rigorous and vigorous quantitative fashion" (92,93). Note that many of these terms have at one time or another been used to describe coatings. A method for describing these terms in a quantitative fashion is becoming a reality. Regarding corrosion, profiles encountered in corrosion pitting have been reported to be similar to those enclosing what are known as Koch Islands. These are mathematical constructions which can be described by fractal dimensions, thus suggesting the application of fractal dimension concepts for description of experimental pit boundaries (91). For general reviews and more details on fractals, see references 75,92-97.

B. Fractal Dimension

The following two paragraphs describing fractal dimension are from Heppenheimer, reference 98. "A fractal dimension is an extension of the concept of the dimension of an ordinary object, such as a square or cube, and it can be calculated the same way. Increase the size of a square by a factor of 2, and the new larger shape contains, effectively, four of the original squares. Its dimension then is found by taking logarithms: dimension = log4/log2 = 2. Hence, a square is two-dimensional. Increase the size of a cube by a factor of 3, and the new cube contains, in effect, 27 of the original cubes; its dimension is log27/log3 = 3. Hence, a cube has three dimensions.

There are shapes-fractals-in which, when increased in size by a factor m, produce a new object that contains n of the original shapes. The fractal dimension, then is log n/log m-evidently the same formula as for squares or cubes. For fractals, for example, in which n = 4 when m = 3, the dimension is log 4/log 3 = 1.26181, A fractal dimension, in short, is given by a decimal fraction; that indeed, is the origin of the term fractal (98)."

The above discussion shows that fractals are expressed not in primary shapes but in algorithms. With command of the fractal language it is possible to describe the shape of a cloud as precisely and simply as an architect might use traditional geometry and blueprints to describe a house (93). A linear algorithm based on only 24 numbers can be used to describe a complex form like a fern. Compare this with the fact that several hundred thousand numerical values would be required to represent the image of the leaf point for point at television image quality (93).

All fractals share one important feature inasmuch as their roughness, complexity or covolutedness can be measured by a fractal dimension. The fractal dimension of a surface corresponds quite closely to our intuitive notion of roughness (97). For example, Figure 34 is a series of scenes with the same 3-D relief but increasing fractal dimension D. This shows surfaces with linearly increasing perceptual roughness: Figure 34(a) shows a flat plane (D ≈ 2.0), (b) countryside (D ≈ 2.1), (c) an old, worn mountain (D ≈ 2.3), (d) a young, rugged mountain range (D ≈ 2.5), and (e) a stalagmite covered plane (D ≈ 2.8).

C. Fractals and Electrodeposition

Fractals could be of importance in the design of efficient electrical. cells for generating electricity from chemical reactions and in the design of electric storage batteries (94). Studies on electrodeposition have become increasingly important since they offer the possibility of referring to a particularly wide variety of aggregation textures ranging from regular dendritic to disorderly fractal (99). The reason electrodeposition is particularly well suited for studies of the transition from directional to "random" growth phenomena is that it allows one to vary independently two parameters, the concentration of metal ions and the cathode potential (74). Much of the interest in this field has been stimulated by the possibilities

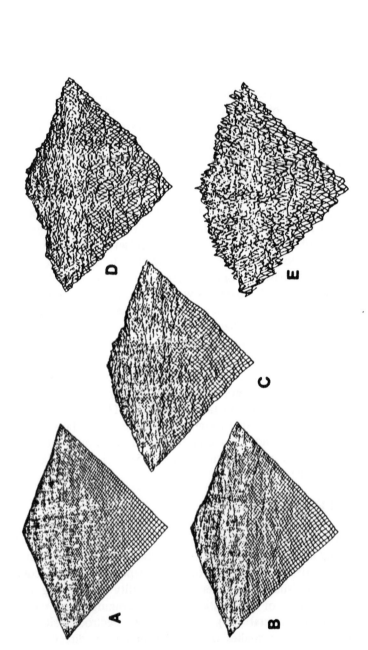

Figure 34: Surfaces of increasing fractal dimension. (a) = 2.0, (b) = 2.1, (c) = 2.3, (d) = 2.5 and (e) = 2.8. From reference 97. Reprinted with permission of IEEE.

furnished by real experiments on electrocrystallization for testing simple and versatile computer routines simulating such growth processes (74,75,- 99-108). As mentioned earlier and shown in Figure 33, a deposit of zinc metal produced in an electrolytic cell has been shown to bear a striking resemblance to a computer generated fractal pattern. The zinc deposit had a fractal measured dimension of 1.7 while the computer generated fractal dimension was 1.71. This agreement is a remarkable instance of universality and scale-invarience. About 50,000 points were used for the computer simulation while the number of zinc atoms, in the deposit is enormously large, almost a billion billion (75).

D. Surface Roughness

Surface roughness is the natural result of acid pickling and abrasive cleaning processes in which, etched irregular impressions or crater like impressions are created in the substrate surface. At present, the effect of surface roughness on the service life of many coating systems is not well understood (109). In some cases, a rough surface may improve the adhesion as discussed in the chapter on Adhesion. In other cases, a rough surface may be detrimental in that it may affect the electrochemical behavior of the surface and make it more difficult to protect the substrate from corrosion. This is due to the fact that a very rough topography requires special care to insure that the peaks of the roughened surface are covered by an adequate coating thickness (109). Surface roughness has to be quantified if one wants to understand its effect on service life. Diamond stylus profilometry is one common method used for this purpose. This technique records a surface profile from which various roughness parameters such as R_a (the arithmetic average roughness) and R_{MAX} (its largest single deviation, can be calculated (110). Although these parameters are widely accepted and used, they are not sufficiently descriptive to correlate surface texture with other surface related measurements such as BET surface area or particle re-entrainment. Fractal analysis has been used to quantify roughness of various surfaces. Figure 35 shows the appearance of surfaces with different fractal numbers. In this example a computationally fast procedure based on fractal analysis techniques for remotely measuring and quantifying the perceived roughness of thermographically imaged, shot, grit and sandblasted surfaces was used (109). The computer generated surfaces compare quite favorably in roughness to the perceived roughness of actual blasted surfaces and provide a three dimensional picture correlating fractal dimension with appearance. Another approach involves application of a fractal determination method to surface profiles to yield fractal-based roughness parameters (110,111). With this technique, roughness is broken

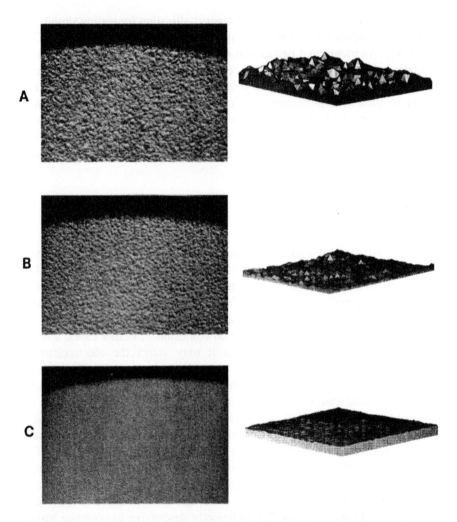

Figure 35: Comparison of surfaces showing blasted steel surfaces on left and computer generated surfaces on right: (A) 5 mil shot blasted surface, D = 2.69, computer D = 2.80; (B) 2.5 mil shot blasted surface, D = 2.48, computer D = 2.50; and (C) 0.5 mil sand blasted surface, D = 2.24, computer D = 2.20. From reference 109. Reprinted with permission of Journal of Coatings Technology.

down into size ranges rather than a single number. This provides a parameter for quantifying the finer structures of a surface. The technique involves use of a Richardson plot and is referred to as "box counting" or the "box" method (110-113). The following description on its implementation is from Chesters et. al. (111). "This method overlays a profilometer curve with a uniform grid or a set of "boxes" of side length b, and a count is made of the non-empty boxes N shown in Figure 36, Then the box size is changed and the count is repeated. Finally, the counts are plotted against each box on a log-log scale to obtain a boxcount plot (Figure 37). The box sizes are back calculated to correspond to the physical heights they would have as features of the profile (hence, the boxcount plot shows counts versus "feature size" rather than box size). It is the absolute value of the slope which gives the fractal dimension, which is referred to as fractal based roughness (Rf). The slope and, hence, Rf will be greater for a rough profile than for a smooth profile."

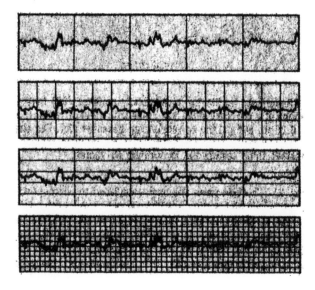

Figure 36: Illustration of box counting as an algorithm to obtain fractal dimension. The rate at which the number of non-empty boxes increases with shrinking box size is a direct measure of fractal roughness. From reference 111. Reprinted with permission of Solid State Technology.

Figure 37 shows box count plots for 316L stainless steel tubing which was given a variety of treatments. Figure 37a is an example of a smooth, electropolished surface. The fractal roughness for the midrange is 1.03 and there is an absence of very small and very large features. Figure 37b is for a chemically polished surface and it is noticeably different from the electropolished surface shown in Figure 37a. It has a unit slope only for features larger than 1 µm and the roughness between 1 and 0.2 µm has two slopes (1.46 and 1.24, respectively). Also, there is a roughness below 0.2 µm (111). Figure 37c , which is for a non-polished surface shows a picture similar to that of Figure 37b except that the roughness between 0.5 and 1.0 µm is higher (110). This analysis provides more information than is obtainable from surface roughness measurements alone, is not limited to profilometers, and can be extended to higher resolution surface techniques (110,111).

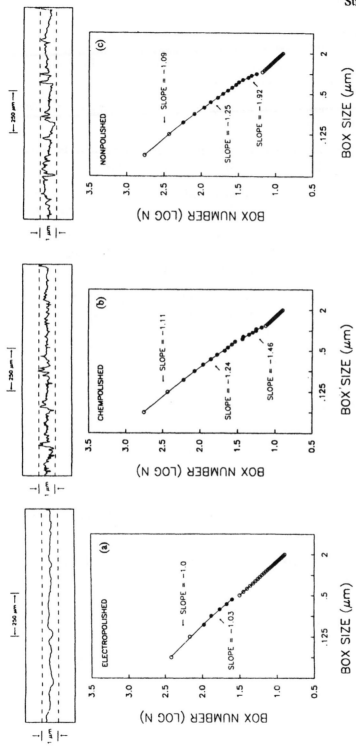

Figure 37: Roughness profiles and box count plots for 316 stainless steel tubing. (a) Electropolished, Ra = 0.05 and Rmax = 0.56; (b) Chemically polished, Ra = 0.09 and Rmax = 1.02; (c) Non-polished, Ra = 0.14 and Rmax = 1.06. From reference 110. Reprinted with permission of Elsevier Science Publishers B. V., The Netherlands.

REFERENCES

1. R. Weil, "Structure, Brightness and Corrosion Resistance of Electrodeposits", *Plating* 61, 654 (1974).

2. N. J. A. Sloane, "The Packing of Spheres", *Scientific American*, 250, 116 (1984).

3. H. McArthur, *Corrosion Prediction and Prevention in Motor Vehicles*, Ellis Horwood Ltd., England, 166 (1988).

4. J. F. Shackelford, *Introduction to Materials Science for Engineers*, *Second Edition*, Macmillan Publishing Company, New York (1988).

5. C. A. Snavely, "A Theory for the Mechanism of Chromium Plating; A Theory for the Physical Characteristics of Chromium Plate", *Trans. Electrochem. Soc.*, 92, 537 (1947).

6. W. H. Safranek, *The Properties of Electrodeposited Metals and Alloys, Second Edition*, American Electroplaters and Surface Finishers Society, Orlando, FL, 1986.

7. V. A. Lamb, "Plating and Coating Methods: Electroplating, Electroforming, and Electroless Deposition", Chapter 32 in *Techniques of Materials Preparation and Handling*, Part 3, R. F. Bunshah, Editor, Interscience Publishers (1968).

8. T. G. Beat, W. K. Kelley and J. W. Dini, "Plating on Molybdenum", *Plating & Surface Finishing*, 75, 71 (Feb 1988).

9. W. Blum and G. B. Hogaboom, *Principles of Electroplating and Electroforming*, Third Edition, McGraw-Hill Book Co., (1949).

10. W. H. Safranek, "Deposit Disparities", *Plating & Surface Finishing*, 75, 10 (June 1988).

11. J. W. Dini, "Deposit Structure", *Plating & Surface Finishing*, 75, 11 (Oct 1988).

12. R. Messier and J. E. Yehoda, "Geometry of Thin-Film Morphology", *J. Appl. Phys.* 58, 3739 (1985).

13. B. A. Movchan and A. V. Demchishin, "Study of Structure and Properties of Thick Vacuum Condensates of Ni, Ti, W, Al_2O_3 and ZrO_2, *Phys. Met. Metallogr.* 28, 83 (1969).

14.   J. A. Thornton, "Influence of Apparatus Geometry and Deposition Conditions on the Structure and Topography of Thick Sputtered Coatings", *J. Vac. Sci. Technol.*, 11, 666 (1974).

15.   J. A. Thornton, "High Rate Thick Film Growth", *Ann. Rev. Mater. Sci.*, 7, 239 (1977).

16.   J. E. Yehoda and R. Messier, "Quantitative Analysis of Thin Film Morphology Evolution", *SPIE Optical Thin Films II: New Developments*, 678, 32 (1986).

17.   R. Messier, A. P. Giri and R. A. Roy, "Revised Structure Zone Model for Thin Film Physical Structure", *J. Vac. Sci. Technol.*, A2, 500 (1984).

18.   J. E. Yehoda and R. Messier, "Are Thin Film Physical Structures Fractals?", *Applications of Surface Science*, 22/23, 590 (1985).

19.   R. Messier, "Toward Quantification of Thin Film Morphology", *J. Vac. Sci. Technol.*, A4, 490 (May/June 1986).

20.   R. Weil, *Electroplating Engineering Handbook*, Fourth Edition, L. J. Durney, Editor, Van Nostrand Reinhold , Chapter 12 (1984).

21.   A. W. Hothersall, "Adhesion of Electrodeposited Nickel to Brass", *J. Electrodepositors' Tech. Soc.*, 7, 115 (1932)

22.   I. Kim and R. Weil, "Tension Testing of Very Thin Electrodeposits", *Testing of Metallic and Inorganic Coatings*, ASTM STP 947, W. B. Harding and G. A. DiBari, Eds., American Society for Testing and Materials, Philadelphia, PA, pp. 11-18 (1987).

23.   R. Haak, C. Ogden and D. Tench, "Comparison of the Tensile Properties of Electrodeposits From Various Acid Copper Sulfate Baths", *Plating & Surface Finishing*, 68, 59 (Oct 1981).

24.   W. D. Fields, R. N. Duncan, and J. R. Zickgraf, "Electroless Nickel Plating", Metals Handbook, Ninth Edition, Vol 5, *Surface Cleaning, Finishing and Coating*, ASM, Metals Park, Ohio (1982).

25.   H. G. Schenzel and H. Kreye, "Improved Corrosion Resistance of Electroless Nickel-Phosphorus Coating", *Plating & Surface Finishing*, 77, 50 (Oct 1980).

26.   "The Engineering Properties of Electroless Nickel Deposits", INCO, New York, NY (1977).

27.   D. H. Johnson, "Thermal Expansion of Electroless Nickel", *Report 28-670, Y-12 Plant*, Oak Ridge National Laboratory, Oak Ridge, TN, (Jan 2, 1982).

28.   D. H. Killpatrick, "Deformation and Cracking of Electroless Nickel", *AFWL-TR-83*, Air Force Weapons Laboratory, Kirtland Air Force Base, NM (Jan 1983).

29.   M. W. Mahoney and P. J. Dynes, "The Effects of Thermal History and Phosphorus Level on the Crystallization Behavior of Electroless Nickel", *Scripta Metallurgia*, 19, 539 (1985).

30.   H. J. Wiesner and W. B. Distler, "Physical and Mechanical Properties of Electroformed Gold-Copper Alloys", *Plating*, 56, 799 (1969).

31.   J. W. Price, "Tin and Tin-Alloy Plating", *Electrochemical Publications Limited* (1983).

32.   C. F. Hornig and J. F. Bohland, "Electrodeposited Tin-Nickel Stability", *Scripta Metallurgia*, 11, 301 (1977).

33.   M. R. Pinnel, "Diffusion Related Behavior of Gold in Thin Film Systems", *Gold Bulletin*, 12, No 2, 62 (April 1979).

34.   C. J. Raub, Proc. Symposium on Economic Use of and Substitution for Precious Metals in the Electronics Industry, *American Electroplaters & Surface Finishers Soc.*, Danvers, Mass. (1980).

35.   S. Jayakrishnan and S. R. Natarajan, "Hydrogen Codeposition in Palladium Plating", *Metal Finishing* 89, 23 (Jan 1991).

36.   S. Nakahara and S. Mahajan, "The Influence of Solution pH on Microstructure of Electrodeposited Cobalt", *J. Electrochemical Soc.*, 127, 283 (1980).

37.   S. Nakahara and E. C. Felder, "Defect Structure In Nickel Electrodeposits", *J. Electrochem. Soc.*, 129, 45 (1982).

38.   R. S. Montgomery and F. K. Sautter, "Factors Influencing the Durability of Chrome Plate", *Wear*, 60, 141 (1980).

39.   C. Conte, G. Devitofrancesco and V. DiCastro, "Study of the Behavior of Protective Metal Surfaces in the Presence of Gas by X-ray, Photoelectron Spectroscopy", *Surface and Coating Technology*, 34, 155 (1988).

40.   T. Barbee, Lawrence Livermore National Laboratory, private communication, May 1989.

41.   H. J. Wiesner and W. J. Frey, "Some Mechanical Properties of Copper Deposited From Pyrophosphate and Sulfate Solutions", *Plating & Surface Finishing*, 66, 51 (Feb 1979).

42.   B. M. Hogan, "Microstructural Stability of Copper Electroplate" *Proceedings SURIFIN 84*, American Electroplaters & Surface Finishers Soc., Orlando, Fl. (1984)

43.   T. T. Saito, "Polishing and Characterizing Diamond Turned Optics". Lawrence Livermore Laboratory, *UCRL 51951*, Nov 3, 1975.

44.   N. P. Fedotov, P. M. Vyacheslavov, and V. I. Gribel, *Journal of Applied Chemistry (USSR)*, 37 (6), 1368 (1964).

45.   J. P. Celis, J. R. Roos and C. Buelens, "Texture in Metallic, Ceramic and Composite Coatings", *Directional Properties of Materials*, H. J. Bunge, Editor, DGM Informationsgesellschaft mbH, Oberursel,Germany, 189 (1988).

46.   S. J. Shaffer, J. W. Morris, Jr., and H.-R. Wenk, "Textural Characterization and its Application on Zinc Electrogalvanized Steels", *Zinc-Based Steel Coating Systems: Metallurgy and Performance*, G. Krauss and D. K. Matlock, Editors, The Minerals, Metals & Materials Society, 129 (1990).

47.   C. S, Barrett, *Structure of Metals*, Second Edition, McGraw-Hill, New York (1952).

48.   J. Amblard, I. Epelboin, M. Froment and G. Maurin, "Inhibition and Nickel Electrocrystallization", *Jour. of Applied Electrochemistry*, 9, 233 (1979).

49.   C. M. Vlad, "Texture and Corrosion Resistance of Metallic Coatings", *Directional Properties of Metallic Coatings*, H. J. Bunge, Editor, DGM Informationsgesellschaft mbH, Oberursel, Germany, 199 (1988).

50.   A. J. Perry and L. Simmen, "Ion-Plated HfN Coatings", *Thin Solid Films*, 118, 271 (1984).

51.   A. J. Perry, "The Structure and Colour of Some Nitride Coatings", *Thin Solid Films*, 135, 73 (1986).

52.   E. Schmid and W. Boas, *Plasticity of Crystals*, English translation published by F. A. Hughes & Co. (1950).

53.   S. J. Shaffer, A. M. Philip and J. W. Morris, Jr., "Micromechanisms of Surface Friction in Zinc Electrogalvanized Steel Sheets", *Proceedings Galvatech 89 Conference*, Tokyo, Japan (Sept 1989).

54.   J. H. Lindsay, R. F. Paluch, H. D. Nine, V. R. Miller and T. J. O'Keefe, "The Interaction Between Electrogalvanized Zinc Deposit Structure and the Forming Properties of Sheet Steel", *Plating & Surface Finishing*, 76, 62 (March 1989).

55.   J. H. Lindsay, "Electrogalvanized Automotive Sheet Steel and the Manufacturing System", Proceedings SUR/FIN 91, *American Electroplaters and Surface Finishers Soc.*, Orlando, Fl., 25 (1991).

56.   G. Krauss and D. Matlock, Editors, *Zinc-Based Steel Coating Systems: Metallurgy and Performance*, The Minerals, Metals & Materials Society (1990).

57.   A. T. Gwathmey, "Effect of Crystal Orientation on Corrosion", *The Corrosion Handbook*, H. H. Uhlig, Editor, Wiley & Sons, (1948).

58.   J. L. Weininger and M. W. Bretter, "Effect of Crystal Structure on the Anodic Oxidation of Nickel", *J. Electrochemical Soc.*, 110, 484 (1963).

59.   G. Raichevski and T. Milusheva, "Effect of the Texture and Structure of the Surface of Electroplated Nickel on its Corrosion and Electrochemical Behavior in an Acid Medium", *Protection of Metals*, 11, No 5, 519 (Sept/Oct 1975).

60.   H. Takechi, M. Matsuo, K. Kawasaki and T. Tamura, *Proceedings 6th Int. Conf. on Textures of Materials*, Vol 2, 209, (1981), Tokyo.

61.   R. F. Ashton and M. T. Hepworth, "Effect of Crystal Orientation on the Anodic Polarization and Passivity of Zinc", *Corrosion*, 24, 50 (1968).

62.   V. Rangarajan, N. M. Giallourakis, D. K. Matlock and G. Krauss, "The Effect of Texture and Microstructure on Deformation of Zinc Coatings", *J. Mater. Shaping Technol.*, 6, 217 (1989).

63.   N. J. Nelson, P. A. Martens, J. F. Battey, H. R. Wenk and Z. Q. Zhong, "The Influence of Texture on the Etching Properties of Copper Foil", *Eighth International Conference on Textures of Materials*, J. S. Kallend and G. Gottstein, Editors, The Metallurgical Soc/AIME, Warrendale, PA., 933 (1988).

64.   J. F. Battey, N. Nelson and P. A. Martens, "Etching Electrodeposited Copper Foil", *PC Fab 10*, 60 (Dec 1987).

65.   I. Artaki, M. H. Papalski and A. L. Moore, "Copper Foil Characterization and Cleanliness Testing", *Plating & Surface Finishing*, 78, 64 (Jan 1991).

66.   H. Leidheiser, Jr., "Coatings", *Corrosion Mechanisms*, F. Mansfeld, Editor, Marcel Dekker, Inc., 165 (1987).

67.   H. Leidheiser, Jr., and D. K. Kim, "Crystallographic Factors Affecting the Adherence of Paint to Deformed Galvanized Steel", *Journal of Metals*, 28, 19 (Nov 1976).

68.   T. F. Davis and D. Kahn, "Contact Resistance of Oriented Electroplated Nickel", Proceedings SUR/FIN 91, *American Electroplaters & Surface Finishers Soc.*, Orlando, FL., 329 (1991).

69.   M. Schlesinger, X. Meng, W. T. Evans, D. A. Saunders and W. P. Kampert, "Micromorphology and Magnetic Studies of Electroless Cobalt/Phosphorus Thin Films", *J. Electrochem. Soc.*, 137, 1706 (1990).

70.   R. J. Morrissey, "The Electrodeposition of Palladium From Chelated Complexes", *Platinum Metals Review*, 27, No. 1, 10 (1983).

71.   M. Kobayashi and Y. Doi, "TiN and TiC Coating on Cemented Carbides by Ion Plating", *Thin Solid Films*, 54, 67 (1978).

72.   J. A. Sue and H. H. Troue, "Effect of Crystallographic Orientation on Erosion Characteristics of Arc Evaporation Titanium Nitride Coating", *Surface and Coating Technology*, 33, 169 (1987).

73.   V. Tomov, D. S. Stoychev and I. B. Vitanova, "Recovery and Recrystallization of Electrodeposited Bright Copper Coatings at Room Temperature. II. X-ray Investigation of Primary Recrystallization", *J. Applied Electrochemistry*, 15, 887 (1985).

74. F. Argoul, A. Arneodo, G. Grasseau, and H. L. Swinney, "Self-Similarity of Diffusion-Limited Aggregates and Electrodeposition Clusters", *Physical Review Letters*, 61, 2558 (1988).

75. L. M. Sander, "Fractal Growth", *Scientific American*, 256, 94 (Jan. 1987).

76. J. McDermott, "Geometrical Forms Known as Fractals Find Sense in Chaos", *Smithsonian*, 14, 110 (Dec 1983).

77. A. L. Goldberger, D. R. Rigney and B. J. West, "Chaos and Fractals In Human Physiology", *Scientific American* 262, 43 (Feb 1990).

78. A. Seri-Levy and D. Avnir, "Fractal Analysis of Surface Geometry Effects on Catalytic Reactions", *Surface Science*, 248, 258 (1991).

79. E. Hornbogen, "Fractals in Microstructure of Metals", *International Materials Reviews*, 34, No 6, 277 (1989).

80. F. F. Ling, "Fractals, Engineering Surfaces, and Tribology", *Wear*, 136, 141 (1990).

81. B. B. Mandelbrot, D. E. Passoja and A. J. Paullay, "Fractal Character of Fracture Surfaces of Metals", *Nature*, 308, 721 (April, 1984).

82. E. E. Underwood and K. Banerji, "Fractals in Fractography", *Materials Science and Engineering*, 80, 1 (1986).

83. C. S. Pande, L. R. Richards and S. Smith, "Fractal Characteristics of Fractured Surfaces", *Jour. of Materials Science Letters*, 6, 295 (1987).

84. Z. H. Huang, J. F. Tian and Z. G. Wang, "Analysis of Fractal Characteristics of Fractured Surfaces by Secondary Electron Line Scattering", *Materials Science and Engineering*, A118, 19 (1989).

85. J. J. Mecholsky, D. E. Passoja, and K. S. Feinberg-Ringel, "Quantitative Analysis of Brittle Fracture Surfaces Using Fractal Geometry", *Jour. American Ceramic Soc.*, 72, (1), 60 (1989).

86. S. Zhang and C. Lung, "Fractal Dimension and Fracture Toughness", *Jour. Phys. D: Applied Physics*, 22, 790 (1989).

87.   K. Ishikawa, "Fractals in Dimple Patterns of Ductile Fracture", *Jour. Materials Science Letters*, 9, 400 (1990).

88.   R. R. Dauskardt, F. Haubensak and R. O. Ritchie,"On the Interpretation of the Fractal Character of Fracture Surfaces", *Acta. Metall.* 38, 143 (1990).

89.   J. R. Ding and B. X. Liu, "Fractal Corrosion of Alloy Films by Acid", *J. Phys.: Condens. Matter*, 2, 1971 (1990).

90.   G. Daccord, "Chemical Dissolution of a Porous Medium by a Reactive Fluid", *Physical Review Letters*, 58, 479 (1987).

91.   J. M. Costa, F. Sagues and M. Vilarrasa, "Fractal Patterns From Corrosion Pitting", *Corrosion Science*, 32, 665 (1991).

92.   B. B. Mandelbrot, *The Fractal Geometry of Nature*, W. H. Freeman & Co., New York (1977).

93.   H. Jurgens, H. O. Peitgen and D. Saupe, "The Language of Fractals", *Scientific American*, 262, 60 (Aug 1990).

94.   B. H. Kaye, *A Random Walk Through Fractal Dimensions*, VCH Publishers, England (1989).

95.   M. LaBrecque, "Fractal Symmetry", *Mosaic*, 16, 14 (Jan/Feb 1985).

96.   M. LaBrecque, "Fractal Applications", *Mosaic*, 17, 34 (Winter 1986/7).

97.   A. P. Pentland, "Fractal Based Description of Natural Scenes", *IEEE Trans. on Pattern Analysis and Machine Intelligence*, Vol. PAMI-6, 661 (Nov 1984).

98.   T. A. Heppenheimer, "Routes to Chaos", *Mosaic*, 17, 2 (Summer 1986).

99.   F. Sagues, F..Mas, M. Vilarrasa and J. M. Costa, "Fractal Electrodeposits of Zinc and Copper", *J. Electroanal. Chem.*, 278, 351 (1990).

100.   R. F. Voss and M. Tomkiewicz, "Computer Simulation of Dendritic Electrodeposition", *J. Electrochem. Soc.*, 132, 371 (1985).

101.   T. Hepel, "Effect of Surface Diffusion In Electrodeposition of Fractal Structures", *J. Electrochem. Soc.*, 134, 2685 (1987).

102.   A. S. Paranjpe, S. Bhakay-Tamhane and M. B. Vasan, "Two Dimensional Fractal Growth by Diffusion Limited Aggregation of Copper", *Physics Letters A*, 140, 193 (Sept. 18, 1989).

103.   C.-Peng Chen and J. Jorne, "Fractal Analysis of Zinc Electrodeposition", *J. Electrochem. Soc.*, 137, 2047 (1990).

104.   G. L. M. K. S. Kahanda and M. Tomkiewicz, "Fractality and Impedance of Electrochemically Grown Silver Deposits", *J. Electrochem. Soc.*, 137, 3423 (1990).

105.   L. Nyikos and T. Pajkossy, "Electrochemistry at Fractal Interfaces: The Coupling of ac and dc Behavior at Irregular Electrodes", *Electrochimica Acta*, 35, 1567 (1990).

106.   J. M. Gomez-Rodriguez and A. M. Baro, "Fractal Characterization of Gold Deposits by Scanning Tunneling Microscopy", *J. Vac. Sci. Technol.*, B9, 495 (Mar/Apr 1991).

107.   V. Fleury, "On A New Example of Ramified Electrodeposits", *J. Mater. Res. 6*, 1169 (1991).

108.   T. A. Witten, Jr., and L. M. Sander, "Diffusion-Limited Aggregation, a Kinetic Critical Phenomenon", *Physical Review Letters*, 47, 1400 (Nov. 9, 1981).

109.   J. W. Martin and D. P. Bentz, "Fractal-Based Description of the Roughness of Blasted Steel Panels", *Journal of Coatings Technology*, 59, No 745, 35 (Feb 1987).

110.   S. Chesters, H. Y. Wen, M. Lundin and G. Kaspar, "Fractal-Based Characterization of Surface Texture", *Applied Surface Science*, 40, 185 (1989).

111.   S. Chesters, H-C. Wang and G. Kasper, "A Fractal-Based Method for Describing Surface Texture", *Solid State Technology*, 34, 73 (Jan 1991).

112.   L. F. Richardson, *General Systems Yearbook*, 6, 139 (1961).

113.   B. B. Mandelbrot, "Fractals in Physics", *Proc. Sixth Trieste International Symposium*, L. Pietronero and E. Tosatti, Editors, Elsevier, New York, 10 (1986).

7
ADDITIVES

INTRODUCTION

The use of additives in aqueous electroplating solutions is extremely important owing mainly to the interesting and important effects produced on the growth and structure of deposits. The potential benefits of additives include: brightening the deposit, reducing grain size, reducing the tendency to tree, increasing the current density range, promoting leveling, changing mechanical and physical properties, reducing stress and reducing pitting. The striking effects on electrocrystallization processes of small concentrations of addition agents, ranging from a few mg/l to a few percent but generally with an effective concentration range of 10^{-4} to 10^{-2} M, point to their adsorption on a high energy surface and deposition on growth sites, thereby producing a poisoning or inhibiting effect on the most active growth sites (1). In fact, as Lashmore has pointed out, "electrodeposition is the science of poisoning; one needs to do something to inhibit the growth of dendrites" (2). The results obtained with additives seem to be out of proportion to their concentration in the solution, one added molecule may affect many thousands of metal ions (1). Their function and mechanism of interaction is not yet clearly understood and their investigation so far has been mostly empirical. Nonetheless, plating additives are extremely important and establishing the proper agents most often determines the success or failure of a given plating process (3).

The generic term "plating additives" covers a wide variety of chemicals which affect deposits in a multitude of ways. The additives can be organic or metallic, ionic or nonionic, and are adsorbed on the plated surface and often incorporated in the deposit (3).

SOME HISTORY AND FOLKLORE

The noticeable influence of additives on deposits coupled with the variety of materials which serve this purpose has led to a type of folklore that of itself is very interesting. Although electrodeposition has progressed from an art to a science, the term black magic is still used in some quarters to describe this technology and additives are one of the key reasons for this description. Figure 1, originally published in 1975 in a review article on electrodeposition shows a magician with his wand and various pots of magic "additives", not atypical for the way many people think about plating. A positive way to look at this is to is that it meets the requirement of Arthur C. Clarke's "Third Law" that "any sufficiently advanced technology is indistinguishable from magic" (4).

Technological innovations and developments generally result from three approaches: 1) application of established or novel scientific

Figure 1: Plating magician with his vats of "magic" additives. From Metal Progress, 107, 71 (Jan 1975). Reprinted with permission of ASM International.

knowledge which leads to the development of a new material or a new device (e.g., the laser), 2) the "trial and error", or Edisonian method and 3) a lucky accident whose potential significance is recognized by the observer (5). The "trail and error" method has been quite fruitful in the field of pharmacology. For example, in one year, the National Cancer Institute spent 75 million dollars testing 30,800 natural and synthetic chemicals as possible agents against cancer. Just 4 of the 30,800 had enough promise to be accepted for human experimentation. An example of the lucky accident approach is the discovery of penicillin by Fleming in 1928. Dust entering through an open window contaminated some cultures he was working on. Fleming observed that areas where the particles had settled resulted in disintegration of bacteria. Penicillin was the resulting discovery leading to the antibiotics revolution of modern medicine (5).

Although there is now a much better understanding of the science governing the mechanisms and functions of additives for plating solutions, much of the work has been of the "trial and error" or Edisonian approach". Many of the situations have been called accidents, at other times, discoveries, but in many cases it was really the intelligent deduction of an observer, upon whom flashed the possibilities of an unexpected happening. Pasteur, who made breakthroughs in chemistry, microbiology, and medicine, recognized this and expressed it succinctly: "In the fields of observation, chance favors only the prepared mind." More recently, Nobel laureate Paul Flory, upon the occasion of receiving the Priestly Medal, the highest honor given by the American Chemical Society, said:

"Significant inventions are not mere accidents. The erroneous view [that they are] is widely held, and it is one that the scientific and technical community, unfortunately, has done little to dispel. Happenstance usually plays a part, to be sure, but there is much more to invention than the popular notion of a bolt out of the blue. Knowledge in depth and in breadth are virtual prerequisites. Unless the mind is thoroughly charged beforehand, the proverbial spark of genius, if it should manifest itself, probably will find nothing to ignite" (6).

It is important to point out that in the past few decades plating research, as well as research in general has changed. Research today is much more based on basic knowledge than on experience since the present day research worker has more fundamental knowledge than those of yesterday and this is shown by the exponential increase in technology (7).

Probably the first addition agent ever used in plating was carbon disulfide in alkaline silver cyanide solutions around 1847 in England (8).

This discovery was not unlike many that were to occur over the years as an observer noticed something different and then pursued the issue with additional experimentation. Wax and resin were used in the process of copying figures for electrotyping. The wax was coated with a film of phosphorus by immersion in a solution of carbon disulfide containing phosphorus. When these wax molds were plated in the cyanide solution, other articles being plated at the same time ended up with a near perfect bright silver deposit. Parts closest to the wax molds were those with the most perfectly bright deposits. This led to the evaluation and then use of carbon disulfide in the plating solution, resulting in the Lyons/Millward patent (8).

Many discoveries of this type followed. As Hendricks has pointed out, the journey from bright plating by means of buffing to bright plating by means of additions agents was an exciting trip that included many happy accidents as well as sound chemical experiments. Interesting reviews covering developments with additives have been provided by Hendricks (9), Soderberg (10) and DuRose (7).

Even Thomas Edison did some research with additives for electrodeposition processes. He used chlorine to obtain improved plating results in nickel solutions (11) and based on this information Henry Brown tried the disinfectant chloramine T, the N-chloro derivative of p-toluene-sulfonamide. This led to the discovery of the use of sulfonamides and saccharin in nickel solutions (7,12).

King discovered bright nickel as a result of his false teeth, to which gum tragacanth had been applied as the adhesive, falling into the plating tank (9,13). A wool sweater which had dropped into an alkaline cadmium plating solution opened the field of bright cadmium plating. It didn't take long before the deposit brightened and this led to Humphries obtaining the first Udylite patent (9,10,14). The stories, documented in the references included in this chapter, get even more folklorish. According to Soderberg (10),the price on the wool additive got so high that alternatives were sought. One installation learned how to produce bright cadmium without it but lost the magic when the "old" foreman quit. This person eventually admitted that he spit in the solution each time he passed by. This led to the discovery that tobacco is a good brightener, although with long use it led to roughness of the deposit. Thus the "Brown Period" of cadmium brightener research was born. Besides tobacco dust, coffee showed promise in cadmium solutions as did Postum and lignin sulfonate wastes (Goulac-which also is a brown powder) from paper mills. Eventually Udylite bought more postum from General Foods than General Foods sold to the public for beverage use (9). Geduld remembered calling on the large cadmium plating department of Chrysler Motors in Detroit in 1953 and seeing mountains of Postum cartons stacked alongside the cadmium plating tanks (15).   Imagine the havoc that would be created in our present environmentally conscience world by storing a food product alongside cyanide plating solutions.

The electroformed copper used for fabricating the combustion chamber of NASA's space shuttle main engine is another interesting story in the history of additives. This deposit serves as a barrier between hydrogen coolant and electrodeposited nickel, preventing embrittlement of the nickel by hydrogen. Jim Pope at the Stanford Linear Accelerator (SLAC) was the first to use this copper since he was asked to provide a deposit that was capable of being brazed in hydrogen without suffering deterioration (16). Electroformed copper deposits contain some oxygen and this can be deleterious when high temperature brazing is done since hydrogen can combine with the oxygen and produce water. With the high temperatures involved, steam pressure generated by the reaction often exceeds the strength of the copper and causes plastic deformation and/or tearing, frequently manifesting itself by grain boundary cracking or cavities (17,18). This phenomenon has been referred to as "hydrogen embrittlement in copper" or "steam embrittlement" (18). Figure 2 in the chapter on Hydrogen Embrittlement shows the effects of this phenomenon on copper.

At SLAC, Pope and his staff experienced much difficulty in attempting to produce a deposit that could be brazed without suffering deterioration. It was brought to Pope's attention that a copper deposit produced in Germany was capable of meeting the brazing requirements. He visited the plant in Germany and discovered that the plating process was quite similar to that used at Stanford. The only noticeable difference was that in Germany, the plating solution was kept in an oak tank while in the US, plastic lined metal tanks were used. With time, the solution used in Germany would leach something from the oak and turn the solution a greenish color unlike the bluish color of copper sulfate solutions. Based on this observation, Pope started experimenting with oak and this led to copper deposits capable of being brazed in hydrogen. SLAC became operational in the early sixties due in no small part to the efforts of Pope and his staff.

When Rocketdyne, NASA's contractor for the space shuttle engine, started working with this copper they knew they had to have a better understanding about its operation other than the simple fact that it leached something from the oak. They challenged their scientists at the Rockwell Science Center in Thousands Oaks, CA (plenty of oak in the area). Eventually it was discovered that a very small amount of sugar was leached from the wood (19). It turns out that all of the pentoses are suitable for use in the copper sulfate solution, e.g. , xylose, arabinose, ribose and lyxose. These materials act as oxygen scavengers in the solution by picking up oxygen. This prevents the anodes from becoming oxidized and leading to this oxygen being incorporated in the deposit.

The previous examples provide some of the interesting history in the complex field of addition agents. Many other examples can be found and more often than not, the common theme is that some very minor ingredient provided the difference between success and failure in regards to appearance and function of the resulting deposit. Some practitioners have claimed that vendors sometimes make subtle changes to their formulations resulting in noticeable changes in the resulting deposit. Case histories presenting this type of information are not available in the technical literature nor will be presented here but it is an important issue to keep in mind when discovering that a process that has worked well for many years changes. Most suppliers of metal finishing products are very diligent about formulation changes and keeping the customer informed. However, to use an example from another field, Ivory Soap, as mature a product as there is, has been reformulated over 80 times (20).

INFLUENCE ON PROPERTIES

The influence of addition agents on physical and mechanical properties of deposits could be the subject of an entire book. In fact, much of the information in Safranek's treatises on properties of electrodeposits relates to this subject (21). Of the multitude of examples available only a few will be presented here. Table 1 from Safranek's book shows the influence of a variety of additives on tensile strength, yield strength and elongation in different nickel plating solutions. Tensile strength is shown to range from 39 to 250 MPa and elongation from 1.0 to 28 percent, depending on the solution and additives used. Another example is Figure 2 which shows the relationship of copper, pH, ammonia, additive, and current density on elongation and tensile strength of deposits produced in pyrophosphate solution (22). Clearly, additive concentration is the most important variable.

An excellent example of the influence of an additive on the properties and microstructure of a deposit is shown in Figure 3 for a copper pyrophosphate solution employing dimercaptothidiazole (DTMD) (28). For the solution without additive, deposits are moderately strong and ductile and have a columnar microstructure (Fig. 3a). Small concentrations of additive enhance nucleation, resulting in a chevron microstructure, high deposit ductility and low tensile strength (Fig. 3b-c). At intermediate levels (above 0.3 to 1.0 cm^3dm^{-3}), unchecked growth of centers nucleated by the additive result in a nodular deposit and decreased deposit ductility and increasing tensile strength (Fig. 3d-f). At higher additive concentrations (1.0 cm^3dm^{-3}), nodule growth is suppressed resulting in fine-grained deposits and both the ductility and tensile strength increase (Fig. 3g-i). At very high additive

Table 1: Effects of Organic Addition Agents on Strength and Ductility of Nickel Deposits*

Agent	Type of Solution	Ultimate Tensile Strength psi	Yield Strength psi	Elongation percent
None	Fluoborate	74,500	52,700	16.6
Saccharin, 1 g/l	Ditto	203,000	120,000	1.0
None	Sulfamate	80,000	50,000	8
Naphthalene trisulfonic acid, 8 g/l	Ditto	140,000	110,000	1.0
Naphthalene trisulfonic acid, 8.7 g/l, and 1 g/l coumarin	Ditto	170,000	140,000	1.0
None	Sulfamate	110,000	70,000	7
Naphthalene trisulfonic acid, 8 g/l	Ditto	150,000	110,000	1.0
Dibenzene sulfonic acid, 1.5 g/l	Sulfamate	75,500	-	12
None	Sulfate	87,500	-	-
Trisulfonated naphthalene, 0.02 g/l	Ditto	250,000	-	-
Trisulfonated naphthalene, 1.0 g/l	Ditto	140,000	-	-
None	Watts	56,000	-	28
Nickel benzene sulfonate, 7.5 g/l, and triamine tolylphenyl methane chloride, 5-10 mg/l	Ditto	212,000	-	5.0

* From Reference 21.

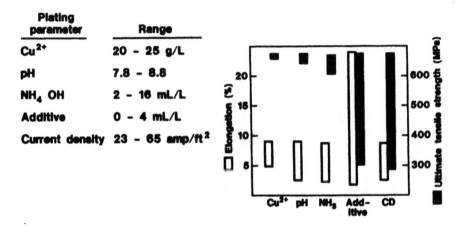

Plating parameter	Range
Cu²⁺	20 - 25 g/L
pH	7.8 - 8.8
NH₄ OH	2 - 16 mL/L
Additive	0 - 4 mL/L
Current density	23 - 65 amp/ft²

Figure 2: Effect of plating parameters on the tensile properties of pyrophosphate copper deposits. Adapted from reference 22.

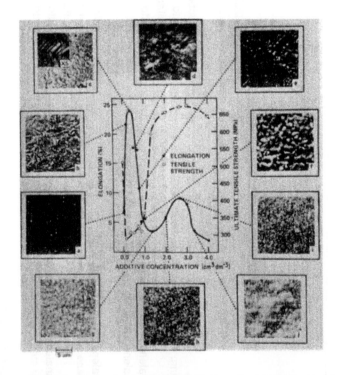

Figure 3: Tensile properties and morphology of annealed pyrophosphate copper deposits versus additive concentration. From reference 28. Reprinted with permission of *The Journal of Applied Electrochemistry*.

levels (3.0 to 4.0 cm^{3}dm^{-3}) the ductility again decreases, presumably because of inclusion of excessive additive in the deposit (Fig. 3j).

INFLUENCE ON LEVELING

Normal electrodeposition accentuates roughness by putting more deposit on the peaks than in the valleys of a plated surface since the current density is highest at the peaks because the electric field strength is greatest in this region. In order to produce a smooth and shiny surface, more metal has to be deposited in the valleys than on the peaks, which is the opposite of the normal effect. The function of certain organic compounds is to produce this leveling in plating solutions. Leveling agents are adsorbed preferentially on the peaks of the substrate and inhibit deposition. This inhibiting power is destroyed on the surface by a chemical reaction which releases it, setting up a concentration gradient close to the surface. An example is coumarin which is used in the deposition of nickel. It adsorbs on depositing nickel by the formation of two carbon-nickel bonds and inhibits nickel deposition probably by a simple blocking action. It is removed from the surface and destroyed by reduction with the main product which is melilotic acid (29).

Radioactive tracer studies have been particularly effective for studying the behavior of addition agents. Additives such as sodium allyl sulfonate, labeled by the reaction between allyl bromide and S labeled sodium sulfite were used in Watts type nickel solutions (30). Grooved brass cathodes were plated with nickel. These substrates had been passivated prior to plating so that the foil could be stripped for counting purposes (Figure 4). Results of the counting experiments (Table 2) show that more activity was deposited on the peaks than in the recesses. Work of this type supports the theory that the addition agent is preferentially adsorbed on the high points of an irregular surface where it acts as an insulator. This inhibits deposition of metal and diverts current to recessed areas (30).

Radioactive tracer techniques, used in Watts nickel solutions, have revealed that a number of mechanisms are feasible, either diffusion and adsorption, or cathodic reduction (31). When two or more compounds were added, the mechanism of incorporation became more complex. Other work on use of radioactive tracer studies with additives can be found in references 32-36.

A practical example of the influence of additives on leveling is shown in Figure 5 (37). A proprietary additive in a copper sulfate solution reduced surface roughness as much as 70 percent with a deposit as thin as 20 um (0. 8 mil). Besides producing deposits which level the hills and valleys on a substrate, levelers also inhibit the formation of asperities such

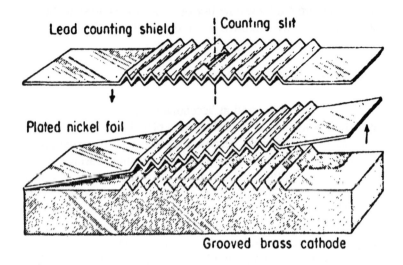

Figure 4: Cathode foil and shield for radiotracer studies. Grooved brass cathodes were plated with nickel which was then passivated to permit stripping of subsequent foils. Counting shield had grooves that limited betas activating the counter to those from either one peak or one valley. Adapted from reference 30.

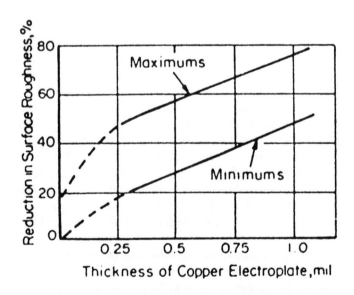

Figure 5: Leveling power of bright copper deposited in copper sulfate solution containing a proprietary additive. Adapted from reference 37.

Table 2:   Lead Slit Counting Rates for Foils Shown in Figure 4*

Foil A -- Top†		Foil B -- Top	
Peak	Recess	Peak	Recess
125	94	42	27
115	72	33	33
115	53	76	53
143	63	47	81
213	80	177	91
226	163	163	160
224	212	94	97
125	65	147	34
106	64	43	28
55	55	59	45
79	57	58	

Foil A -- Bottom		Foil B -- Bottom	
Obverse of peak	Obverse of recess	Obverse of peak	Obverse of recess
248	143	145	58
198	103	133	77
152	133	134	84
120	112	154	65
207	102	135	100
254	113	190	128
227	163	212	126
150	146	176	101
146	129	163	82
181	158	144	80
242	129	102	

* Counting was left to right on the top of the foil and right to left on the bottom, so that the values in the columns are matched.

† "Top" refers to the side next to the solution during plating, "bottom" to the side next to the cathode.

* From Reference 30.

as nodules.   This increases the stability of the deposition process, particularly for thick coatings (29).

INFLUENCE ON BRIGHTENING

A bright deposit is one that has a high degree of specular reflection (e.g., a mirror), in the as-plated condition.   Although brightening and leveling are closely related, many solutions capable of producing bright deposits have no leveling ability (38).   If the substrate is bright prior to plating, almost any deposit plated on it will be bright if it is thin enough. However, a truly bright deposit will be bright over a matte substrate and it will remain bright even when it is thick enough to hide the substrate completely.   Plating solutions without addition agents seldom or never produce bright deposits.

There is a direct relationship between brightness and surface structure of electrodeposits as shown in Figure 6 (39).   The measure of smoothness used in this example is the fraction of the surface area which does not deviate from a plane by more than 0. 15 µm, which is of the order of the wavelength of visible light.   This value was chosen because it has been found that with specularly bright nickel, there are no hills higher or valleys deeper than 0. 15 µm (39,40).

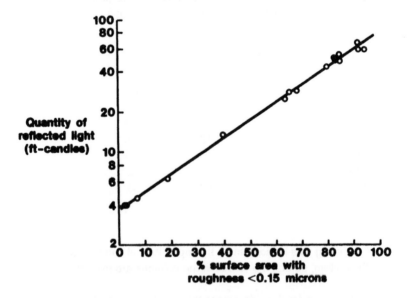

Figure 6: Relationship between quantity of reflected light (brightness) and fraction of area with roughness less than 0.15 um. Adapted from reference 40.

CLASSIFICATION AND TYPES OF ADDITIVES

Additives can be classified into four major categories: 1-grain refiners, 2-dendrite and roughness inhibitors, 3-leveling agents, and 4-wetting agents or surfactants (3). Typical grain refiners are cobalt or nickel codeposited in trace amounts in gold deposits. Dendrite and roughness inhibitors adsorb on the surface and cover it with a thin layer which serves to inhibit the growth of dendrite precursors. This category includes both organic and inorganic materials with the latter typically being more stable. Leveling agents, such as coumarin or butynediol in nickel solutions, improve the throwing power of the plating solution mostly by increasing the slope of the activation potential curve. The prevention of pits or pores in the deposit is the main purpose of wetting agents or surfactants (3).

Metals differ in their susceptibility to the effect of additives, and the order of this susceptibility is roughly the same as the order of their melting points, hardness and strength; it increases in the order Pb, Sn, Ag, Cd, Zn, Cu, Fe, Ni (41). Thousands of compounds are known that brighten nickel deposits from the sulfate-chloride solution, while it is only fairly recently that ways of brightening tin deposits from acid solutions have been developed. The progression in the series corresponds to: 1) the increasing tendency of metal ions to form complexes and 2) to increasing activation polarization from simple ions. This is in the reverse order to the overvoltages observed in the evolution of hydrogen on metal cathodes. Lyons suggests that: "An atom which is capable of interacting strongly with other atoms of the same or other kinds tends to form a strong crystal lattice with a relatively high melting point, to coordinate strongly with ligands, to decoordinate water slowly, and to catalyze conversion of atomic to molecular hydrogen" (41).

Additives are often high molecular weight organic compounds or colloids since small ions or molecules are generally not very effective (42). This is shown in Table 3 which relates minimum concentration of organic compounds required to impart appreciable brightness to nickel deposits (43). The size of the molecule can also influence the stress in the deposit. Coumarin, which is a small molecule compared to phenosafranine (Figure 7) reduces macrostress in nickel deposits, whereas, phenosafranine increases tensile macrostress (44).

An open discussion of the components of brightener systems is difficult because many of these systems are proprietary. Suppliers guard their formulations from distribution simply because the brightener market is so competitive. However, there are numerous technical publications detailing many of the additives commonly used. A listing of the materials that have been used as additives in plating solutions culled from the open

Table 3: Relationship Between Molecular Size and Minimum Concentration of Organic Compound Required to Cause Brightening in Nickel Plating*

Type of Compound	Example	Avg. Min. Conc. to Brighten Deposit (ml/l)
Very large	Magenta dye	0.000057
Bicyclic	Saccharin	0.00085
Monocyclic	Furfural	0.008
Short chain alkyl compounds	Acryonitrile	0.003

* From Reference 43.

literature would be monumental and will not be attempted here. However, some limited examples will be presented in the material that follows.

Cadmium — Glue was used in solutions for electrowinning of cadmium from around 1910 (45). The first bright cadmium plating solution was introduced in 1925 and consisted of a cyanide solution plus a caustic solution of proteins (14). Some of the addition agents that have been used in the ensuing years for cadmium include sulphonated castor oil (Turkey Red Oil), aromatic aldehydes, and inorganic salts such as nickel or cobalt compounds (46).

Copper — Some of the materials that have been used with acid copper include glue, dextrose, phenolsulfonic acid, molasses and thiourea. Many of the present day commercially available brighteners contain three components designated as carrier, leveler and brightener. Reid suggests that: "Carriers are typically polyalkylene glycol type polymers with a molecular weight around 2000, levelers are typically alkane surfactants containing sulfonic acid and amine or amide functionalities, and brighteners are typically propane sulfonic acids which are derivatized with surface active groups containing pendant sulfur atoms" (47).

Additives for cyanide copper systems include compounds having active sulfur groups and/or containing metalloids such as selenium or tellurium. Other agents that have worked are organic amines or their reaction products with active sulfur containing compounds; inorganic compounds containing such metals as selenium, tellurium, lead, thallium, antimony, arsenic; and organic nitrogen and sulfur heterocyclic compounds (48). An extensive listing of additives used in acid and cyanide copper prior to 1959 can be found in reference 48.

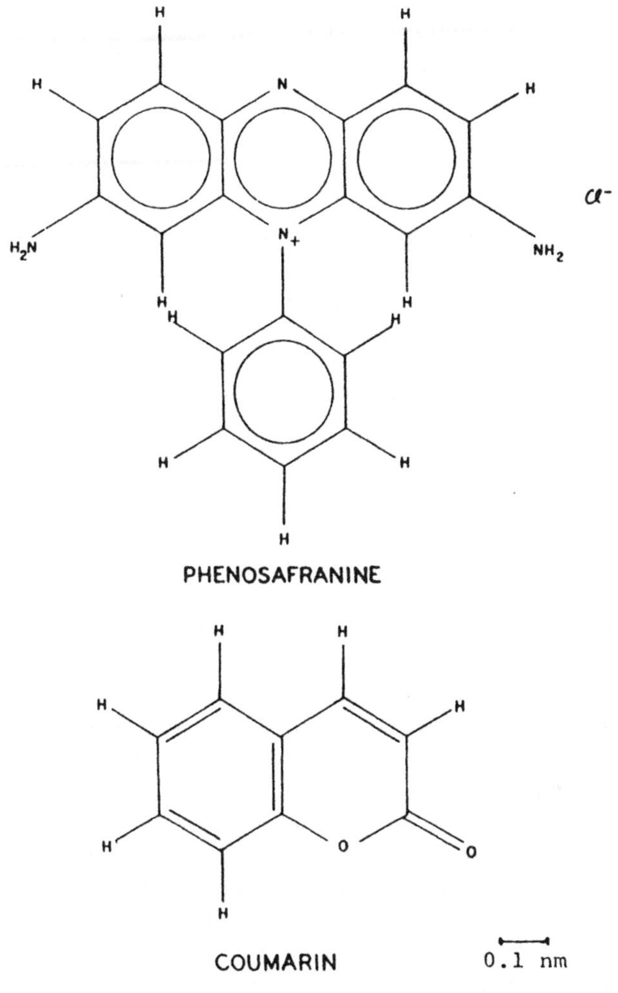

PHENOSAFRANINE

COUMARIN   0.1 nm

Figure 7:   Schematic representation of coumarin and phenosafranine molecules drawn approximately to scale.   From reference 44.   Reprinted with permission of The Electrochemical Soc.

Gold — There are three principal types of additives associated with high purity gold electrolytes: complexing agents, grain refiners, and hardening agents (49).   Complexing agents such as pyrophosphate ion, organophosphorus compounds and polyphosphates are added to reduce the activity of metallic impurities in the solution by forming stable complexes and hence minimizing codeposition.   Organic chelating agents such as EDTA and related compounds are also used.   Relatively small amounts of base metals are used for providing grain refinement.   These additives also provide smoothing and semi-brightening of the deposit,while not being codeposited to a significant extent. In neutral solutions, arsenic and thallium have been used. Some additives such as alums and hydrazine sulfate have been claimed to harden the electrodeposit without being codeposited (49).

Use of heavy metal ions in trace quantities (parts per million) in gold electroplating solutions, induces a marked cathodic depolarization which extends the range of current densities over which smooth, fine grained deposits can be obtained (50-52). In slightly alkaline phosphate electrolytes the most effective additives comprise the family of elements Hg, Tl, Pb and Bi, which lie immediately adjacent to gold in the periodic table. They exhibit a strong tendency to form an adsorbed monolayer on gold and platinum electrodes. This is done at potentials positive to those at which their cathodic deposition as bulk metals would begin, i.e. at underpotentials (52). Deposits obtained with these additives have a very fine and highly uniform grain size (51). The various heavy metal ions have a brightening effectiveness which is in the order Tl > Pb > Bi > Hg. This is the inverse order of their electron work functions (the amount of energy required to lift an electron out of a lattice). The postulated mechanism for this performance is that the elements form an adsorbed monolayer on the surface of the gold. This lowers its work function and thereby lowers its deposition potential so that deposition then occurs at underpotentials (52). An excellent review on additives for gold plating systems can be found in reference 53.

Lead — Common additives for deposition of lead from fluoborate solutions include peptone and resorcinol (54). For plating strip, which requires high current densities of 1000 amp/ft^2 or greater, hydroquinone was the best additive out of 230 compounds evaluated on the basis of performance, stability, cost and lack of industrial hazard (55). Compounds which provided grain refinement and 1000 amp/ft^2 or better,limiting current density were the following structural groups listed in decreasing order of effectiveness: aliphatic compounds, benzene derivatives, naphthalene derivatives, anthraquinone derivatives and heterocyclic compounds.

Nickel — The key to modern bright nickel plating was the discovery of combining an organic "carrier" brightener with an auxiliary compound to produce brightness and leveling (45). These are referred to as Class 1 and Class II brighteners and materials of each type are listed in Table 4. Brighteners of the first class have two functions: 1-provide bright deposits over a bright substrate and 2-permit the second class brighteners to be present over an acceptably wide range of concentrations. Brighteners of the second class are used to build mirror-like lustre. However, most of these lead to excessive brittleness and stress in deposits in the absence of brighteners of the first class (56). Comparisons of the two brightener

Table 4: Brighteners For Nickel (Reference 56)

Brighteners of the First Class

Source of =C—	Name
1. Aryl ring	Benzene, naphthalene, etc.
2. Substituted aryl ring	Toluene, xylene, naphthylamine, toluidine, benzyl naphthalene, etc.
3. Alkylene chain	Vinyl, allyl, etc.

Linkage with —SO₂=

1. —OH	Sulfonic acid
2. —ONa; —O\Ni/—O	Sulfonates
3. —NH₂	Sulfonamides
4. \NH/	Sulfonimides
5. —H	Sulfinic acid
6. —R (organic radical)	Sulfones

Brighteners of the Second Class

Period	GROUP				
	IIB	IIIA	IVA	VA	VIA
3	—	—	—	—	S
4	Zn	—	—	As	Se
5	Cd	—	(Sn)	(Sb)	Te
6	Hg	Tl	Pb	(Bi)	—

( ) denotes metal not commonly used.

Radical	Type
C=O	Carbon monoxide, ketones, aldehydes, carbolic acids, proteins
C=C	Alkylenic carboxylic esters, alkylenic aldehydes, aryl aldehydes, sulfonated aryl aldehydes, allyl, vinyl, etc. compounds, coumarin.
C≡C	Acetylene and derivatives, acetylenic alcohols
C—N	Azine, thiazine, and oxazine dyes, triphenyl methane dyes, quinidines, pyrimidines, pyrazoles, indazoles, pyridinium and quinolinium compounds
C≡N	Ethylene cyanohydrin
N—C=S	Thiourea, cyclic thioureides
N=N	Azo dyes

Table 5: Comparison of Carriers and Brighteners Used In Nickel Plating Solutions*

Carriers (Class I)	Brighteners (Class II)
Bright or cloudy deposits, unable to provide high lustre with continued plating	Brilliant leveling and increasing lustre
Sulfur (0.03%) occluded in deposit when Class I compounds are used without Class II compounds	Introduces carbon in the deposit
No critical upper concentration, used in high concentrations(I-I0 g/l). Cathode potential increases 15-45 mv in low concentrations, then very little change with further additions.	Cathode potential continues to increase sharply with increases in concentration
Do not cause cracking or peeling.	Deposits crack and peel when the cathode potential increase exceeds approximately 30 mv
Reduce stress, can result in compressive stress. Lessen ductility slightly.	Have a very deleterious effect on properties, producing brittle, highly stressed deposits.

* From references 43 and 59.

classes are provided in Table 5. For more detail on nickel plating brighteners, see references 43, 45, 46, 56-59 and the Corrosion chapter in this book.

Silver — Present silver solutions closely resemble the one described in the first patent over 140 years ago (8). Carbon disulfide and thiosulfate have been the most widely used addition agents over the years. Many other materials have been proposed, including gums, sugars, unsaturated alcohols, sulfonated aliphatic acids, xanothogenates, Turkey Red Oil, Rhodamine Red and compounds of antimony and bismuth (46,60). Most of these agents are sulfur bearing organic compounds or reaction products of sulfur and organic compounds (60).

Tin — Similar to most acid solutions, deposits of tin from acid solutions containing no additives are crystalline and nonadherent. The development of smooth deposits free from treeing resulted from years of extensive research on a long list of addition agents. DuRose (61) reviewed the types of addition agents reported in the literature to about 1960 and MacIntosh (62) and Dennis (46) to the early 1970's.

Tin-Lead — Some of the common additives are glue, resorcinol, nicotine, peptone, beta-naphthol, biphenyl sulfones and ethoxy ethers (63). Coatings of terne alloy containing up to 14% tin have been electrodeposited from fluoborate solutions containing hydroquinone as the addition agent (64).

Zinc — Typical additives from research in the early 1900's for acid zinc plating included dextrose, dextrin, glucose, beta naphthol, vegetable gums, gelatin, brewers yeast and licorice (15, 65, 66). An excellent review of the early history of additives for zinc plating as well as development of brighteners for the various zinc systems has been provided by Geduld (15). Two early zinc plating additives which did not amount to much commercially but which paved the way for future developments were naphthalene disulfonate (67) and pyridine (68). These opened up an area of investigation into the use of heterocyclic organic additives in zinc plating which eventually led to the primary constituents of many brighteners used in the field today, especially in bright cyanide zinc plating (15).

Compounds used as brighteners in zinc cyanide solutions include aromatic aldehydes such as anisaldehyde, polyvinyl alcohol, glue, gelatin, and sodium sulfide (46). Crotty reviewed the patent literature and pieced together skeletal brightener formulations to illustrate the functional properties of a system, demonstrating the roles of carriers and brighteners in alkaline cyanide, non-cyanide and acid chloride systems (69). Earlier, DaFonte provided detailed discussion of additive chemistry for zinc plating systems (70). Zinc plating additives can be broken down into three categories: carriers, brighteners and purifiers. Carriers provide a smooth deposit and also prevent the formation of dendrites. The brighteners form a truly bright deposit by adding clarity to the smoky deposit provided by the carrier. Purifiers are used to remove the last traces of smokiness in the deposit. This smokiness might result from impurities in chemicals from the plating solution or from metals that are present in zinc anodes (69).

MECHANISMS

Additives act as grain refiners and levelers because of their effects on 1) electrode kinetics and 2) the structure of the electrical double layer at the plating surface (71). Since additives are typically present in extremely small concentrations, their transport towards the electrode is nearly always under diffusion control and, therefore, quite sensitive to flow variations (3). The effects of additives are often manifested by changes in the polarization characteristics of the cathode. Many are thought to function by adsorption on the substrate or by forming complexes with the metal. This results in development of a cathodic overpotential which is maintained at a level which allows the production of smooth, non-dendritic plates having the desired grain structure (71). An example is bright nickel deposition which is accompanied by a cathode potential increase (polarization) of the order of 20 mv, or more as shown in Figure 8 (43).

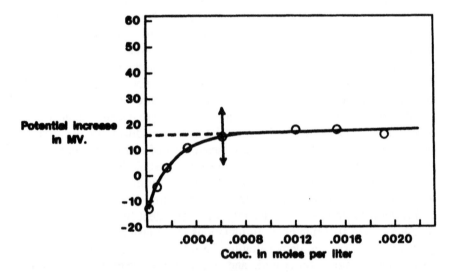

Figure 8: Cathode potential-concentration curve for 1 naphthylamine 4, 8 disulfonic acid. The first sign of brightening of the nickel deposit is indicated by an arrow. Adapted from reference 43.

Numerous mechanisms have been suggested to explain behavior of additives: 1) blocking the surface, 2) changes in Helmholtz potential, 3) complex formation including induced adsorption and ion bridging, 4) ion pairing, 5) changes in interfacial tension and filming of the electrode, 6) hydrogen evolution effects, 7) hydrogen absorption, 8) anomalous codeposition, and 9) the effect on intermediates. These are discussed in detail in a comprehensive review by Franklin (72). Additional excellent coverage on mechanisms of levelling and brightening of addition agents can be found in the recent paper by Oniciu and Muresan (72a).

DECOMPOSITION OF ADDITION AGENTS

Addition agents are generally consumed in the deposition process. For example, in the case of nickel they may be decomposed and the products in part incorporated in the deposit (sulfur, carbon, or both) or released back into the electrolyte. At a pH of 4, approximately 90% of the coumarin consumed at the cathode is reduced to melilotic acid and incorporated in the deposit (46). Radiotracer work has shown virtually complete molecules of melilotic acid, of approximately 10Å incorporated in nickel deposits plated from solutions containing coumarin (31).

Figure 9 relates labeled sulfur content of a nickel deposit to concentration of saccharin in the solution and is similar in shape to that obtained with carbon when coumarin is used in nickel solutions. Breakdown products of additives can affect internal stress in the deposit. For example, eight decomposition products are possible with saccharin (benzoic acid sulfimide) and these are listed in Figure 10. Of these eight products, it has been shown that o-toluene sulfonamide and benzamide are found in Watts nickel solutions when saccharin is used (73). Figure 11 shows the build up of these two decomposition products as a function of solution electrolysis time and their influence on stress in the deposit. In the case of these products, when their concentration gets too high, they are removed by treatment with activated carbon.

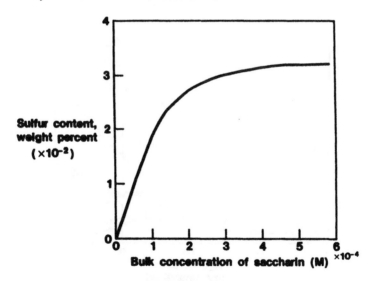

Figure 9: Relationship between bulk concentration of saccharin in a Watts nickel plating solution and sulfur content of deposit. Plating conditions: temperature 55°C, pH 4.4 and current density 4 A/dm². Adapted from reference 46.

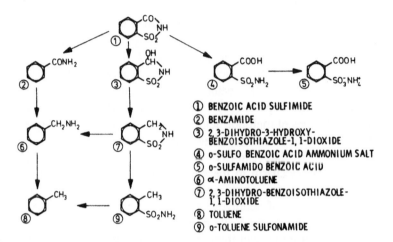

① BENZOIC ACID SULFIMIDE
② BENZAMIDE
③ 2,3-DIHYDRO-3-HYDROXY-
   BENZOISOTHIAZOLE-1, 1-DIOXIDE
④ o-SULFO BENZOIC ACID AMMONIUM SALT
⑤ o-SULFAMIDO BENZOIC ACID
⑥ α-AMINOTOLUENE
⑦ 2,3-DIHYDRO-BENZOISOTHIAZOLE-
   1, 1-DIOXIDE
⑧ TOLUENE
⑨ o-TOLUENE SULFONAMIDE

Figure 10: Possibilities for the decomposition of benzoic acid sulfimide (saccharin). From reference 73. Reprinted with permission of The American Electroplaters & Surface Finishers Soc.

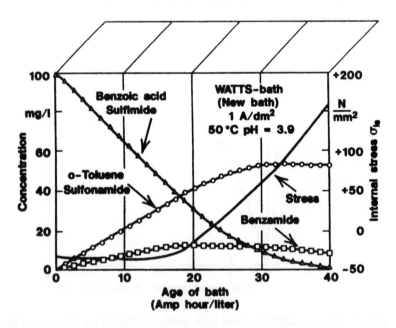

Figure 11: Decomposition of benzoic acid sulfimide (saccharin) during electrolysis and its influence on stress. From reference 73. Adapted from reference 73.

CONTROL AND ANALYSIS OF ADDITIVES

Lack of control of plating solutions is a major problem which leads to reduced reliability and increased costs for plated parts. One reason that progress in this area has been slow is the difficulty of performing quantitative analysis on the additives, often a mixture of two or more compounds (not to mention the numerous additive breakdown products that can accrue with time) in the ppm and ppb ranges in the presence of high concentrations of electrolytes (74). Techniques that are available include the Hull cell, bent cathode, chromatography, a variety of electroanalytical methods,impedance probes and spectrophotometry.

HULL CELL

A number of researchers have stated that the Hull Cell has probably contributed more to the advancement of electroplating than any other tool (15,75) and this is likely true. Jackson and Swalheim contend that "a plater without a Hull Cell is like an electrician without a voltmeter" (76). It is a simple, easy test to run and does not require advanced technical training for interpretation of results.   With this test one can determine plating characteristics over at least a tenfold change in current density range. Although it is not nearly as exotic and complex as many of the other analytical tools discussed in this section the fact that this tool, first demonstrated in 1939 (77) is still viable shows that it has weathered the test of time (78).

The Hull Cell is a trapezoidal box of non-conducting material with one side at a 37.5 degree angle (Figure 12).   An anode is laid against the right angle side and a cathode panel is laid against the sloping side.   When a current is passed through the solution sample contained in the cell, the current density along the sloping cathode varies in a known manner.   In this way the character of the deposit at a range of current densities is determined in one experiment.   Current used in the cell varies from 1 to 3 amps, and time from 2 to 10 minutes, depending on the type of solution being tested. Special rulers or scales are available that are marked to show specific current densities on a plated Hull Cell panel depending on the amperage used (78) (Figure 13). The standard Hull Cell is 267 ml capacity, a volume selected in premetric days because 2 grams of material added to the cell corresponds to a 1.0 oz/gal addition to the main plating solution. Today Hull Cells are available in a variety of sizes including 500 ml and 1 liter to fulfill needs of those working in the metric system.

A Hull cell panel gives more information than the useful plating range. It reveals a pattern of bright, semi-bright, dull, burned, pitted and

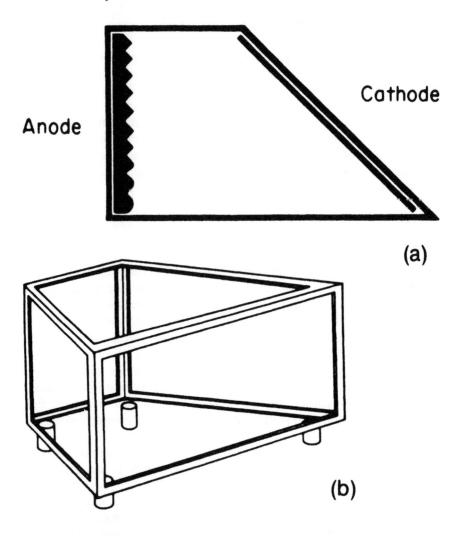

Cathode

Anode

(a)

(b)

Figure 12: Hull cell: a) top view, b) plan view of 267 ml cell.

cracked areas that typically describe results of a specific test. Figure 14 shows one technique for recording data from Hull Cell panels (79).

Modifications to the Hull Cell have appeared including a cell with holes in the two parallel sides to permit solution circulation while the cell is immersed in the actual plating tank under evaluation (80). . This cell can be operated for long periods of time without temperature fluctuation. A more recent modification is the Gornall Cell (Figure 15) which is used for testing solutions for the printed wiring board industry (81). This version allows for plating of samples with drilled holes and provides accurate

Current density, A/ft²

(Move decimal point one place to left for A/dm²)

1 Ampere (A)	40	30	25	20	15	12	10	8	6	4	3	2	1	0 5
2 A	80	60	50	40	30	24	20	16	12	8	6	4	2	1
3 A	120	90	75	60	45	36	30	24	18	12	9	6	3	1 5
5 A	200	150	125	100	75	60	50	40	30	20	15	10	5	2 5

267- and 534-mL Hull Cells

Figure 13: Hull cell ruler. Adapted from reference 78. Note: This is not to scale since it was reduced for printing purposes.

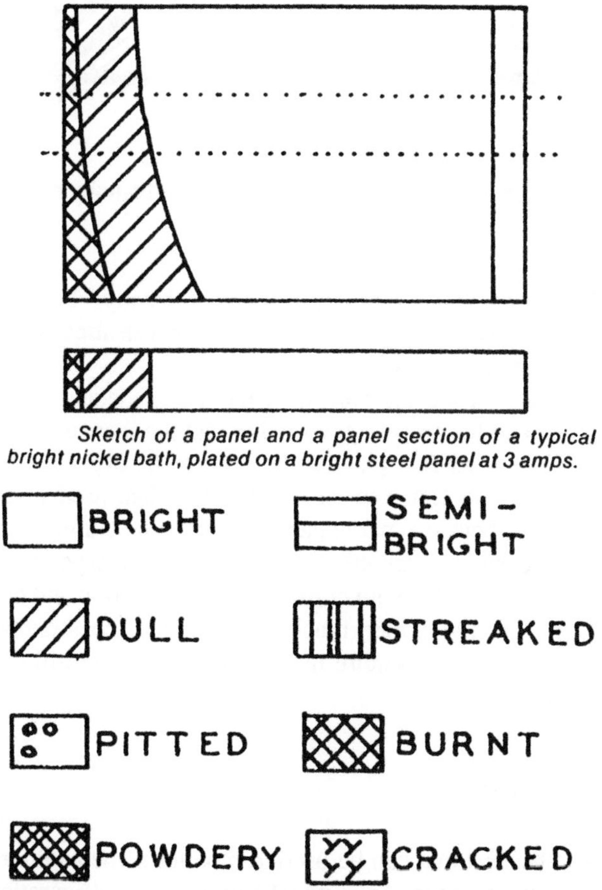

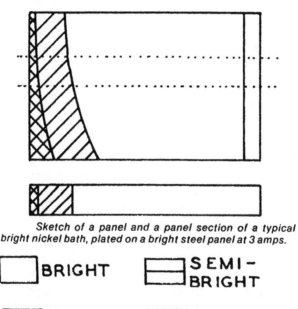

Sketch of a panel and a panel section of a typical bright nickel bath, plated on a bright steel panel at 3 amps.

BRIGHT   SEMI-BRIGHT

DULL   STREAKED

PITTED   BURNT

POWDERY   CRACKED

Figure 14: Code defining the surface appearance of a Hull cell planel. From reference 79. Reprinted with permission of Metal Finishing.

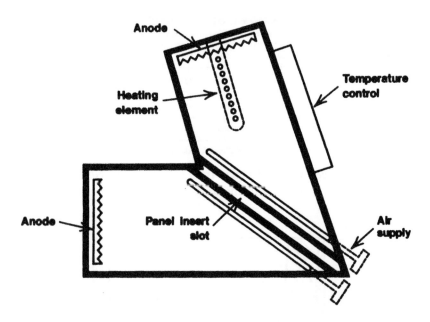

Figure 15: Gornall cell for studying surface-to-hole ratios for printed wiring boards. Adapted from reference 81.

surface-to-hole ratios as a function of current density. For more detail on the Hull Cell and its operation besides the references already cited, the book by Nohse (82) is recommended.

However, while the Hull Cell has been adequate in the past, the increasing demand for process automation and the increasing complexity of parts makes the need for quantitative control of organic additives more important. The Hull Cell can be misleading when used to evaluate high speed processes because parameters related to solution flow, solution geometry and current distribution are not always reproduced (83). Also, new additives required for high current density operation and other advances will have to function under much more severe constraints than those presently in use and will require more sophisticated control than attainable with the Hull Cell.

BENT CATHODE

Another test that can be used to monitor some additives is the bent cathode. Shown in Figure 16, a panel bent in this configuration and plated in either a beaker in the laboratory or in the actual production tank can provide information on leveling, burning and striations, roughness, low

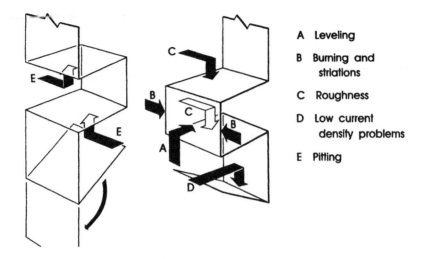

A   Leveling

B   Burning and
     striations

C   Roughness

D   Low current
     density problems

E   Pitting

Figure 16:   Bent cathode test panel. Adapted from reference 84.

current density problems and pitting. Inspection is performed after opening the bottom portion of the panel. Details on operation of this test and its value for controlling acid copper sulfate solutions can be found in reference 84.

CHROMATOGRAPHY

Chromatographic techniques have been finding increased usage in monitoring of plating solutions (85). Their main attraction is the ability to quantitate simultaneously low concentrations of several inorganic and organic solution constituents in one analytical run. With chromatography, the usage of two terms-High Performance Liquid Chromatography (HPLC) and Ion Chromatography (IC) has caused some confusion. For this reason a more general term, Liquid Chromatography (LC) is now preferred (85). A detailed explanation of modern LC methods can be found in reference 86 and applications of LC for analysis of electroplating additives in Table 6.

Liquid chromatography is shown schematically in Figure 17. It consists of four modes: 1) a sample delivery mode — this includes a high-pressure analytical pump and a sample injection valve; 2) a separation mode — this includes one of a variety of analytical separator columns; 3) a detection mode — this includes one of a variety of detectors; and 4) a data reduction mode — this can range from a simple X-Y strip chart recorder to a personal computer system (87).

Table 6: Plating Solution Additives Analyzed by Various Techniques

Solution	Additive	Analysis Method*	Reference
Copper			
Acid sulfate	Proprietary	V	104,108,109,110 111,112
	"	C	47,85,87,99,113-115
	"	I	105
	Thiocarbamoyl-thio- alkane sulfonates	V	116
	Mercaptopropane sulfonic acid	C	117
	Thiourea	P	118
	"	I	74,119
	Polyethylene glycol	I	120
	Polyether sulfides	V	96
	Sulfoniumalkane sulfonates	V	97
	N,N-dimethylaniline	I	74
	2-mercaptobenzothiazole	I	74
Pyrophosphate	Proprietary	V	121
	Dimercaptothiadiazoles	V	92-95,98,101,102
Electroless	Adenine, guanine. saccharin, coumarin	V	122
	Mercaptobenzothiazole	V	83,123
	"	P	118
Gold			
Acid cyanide	Co and Ni hardeners	P	124
Lead			
Fluoborate	Lignin sulfonate	V	71
Perchlorate	Rhodamine-B	V	125

(continued)

Table 6: (continued)

Nickel	Proprietary	C	85
	"	P	129,130
	Pyridine derivatives	V,S	107
	"	P	126
	Saccharin	V,S	107
	"	C	73,85,88,127,128
	"	P	118
	Acetylenic alcohols, aromatic sulfonamides	C	130
	Wetting agents	C	128
	O-benzaldehyde, sulfonic acid	P	118
	2 butyne 1, 4 diol	C	85
	"	I	131
	Rhodamine B,sodium saccharin	V	106
	Sodium benzene-sulfonate	I	131
	Sodium lauryl sulfate	C	85,128
Palladium	Hydroquinone	P	118
Silver Cyanide	Propargyl alcohol,2,5-dimethyl-2,5 hexane diol	I	132
Tin	Proprietary	C	88,133,134
	"	P	136
Tin-lead Fluoborate	Proprietary	V,P	135,110
	Resorcinol	C	127
Zinc	Benzoic acid,anisal-dehyde,vanillin, and many other compounds O-chlorobenzaldehyde	C	69,130
		P	90, 118

* V=voltammetry
   C=chromatography
   P=polarography
   I=impedance
   S=spectrophotometry

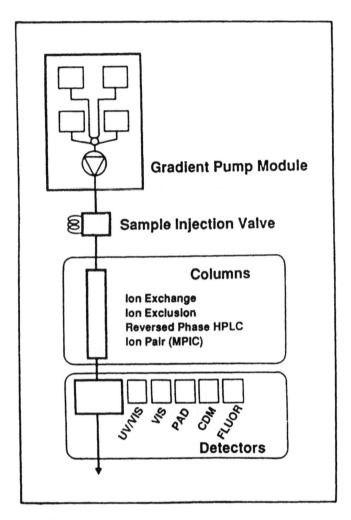

Figure 17: Schematic of an HPLC system. Adapted from reference 87.

As shown in Table 6, chromatographic techniques have been used to analyze additives in acid copper, nickel, tin, tin-lead and zinc plating solutions. Figure 18 shows an HPLC chromatogram of some organic brighteners cited in the literature for zinc plating (69). The breakdown products of saccharin discussed earlier and shown in Figures 10 and 11 were determined by chromatography.

A recent innovation includes the use of scanning UV detectors in the chromatographic system. Unlike conventional single wavelength two dimensional chromatograms, the recordings produced by scanning detectors are three dimensional. These 3-D chromatographic plots facilitate the

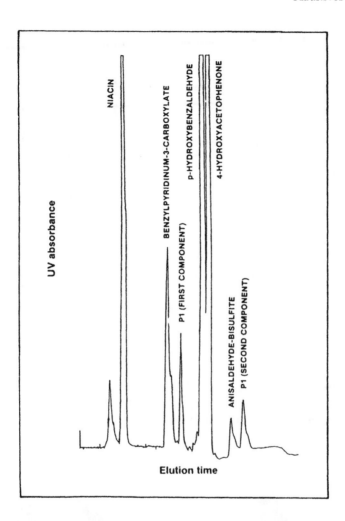

Figure 18: HPLC chromatogram of some organic brighteners cited in the literature for zinc plating. From reference 69. Reprinted with permission of the American Electroplaters & Surface Finishers Soc.

identification of unknown peaks and help to improve knowledge of chemical reactions responsible for the properties of deposits. Figures 19 and 20 show respectively a conventional chromatogram and the corresponding 3-D plot obtained. The peaks 1,2,3 and 4 came from one additive solution and the two late eluting peaks 5 and 6 were attributed to a replenisher solution. Peak 2 originating from the replenisher additive could be identified as phenol which was a degradation product from the organic additive (88).

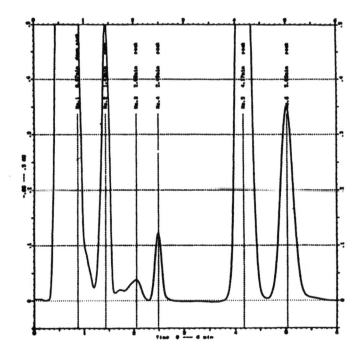

Figure 19: Single wavelength (210 nm) chromatogram of a tin plating solution. From reference 88. Reprinted with permission of the American Electroplaters & Surface Finishers Soc.

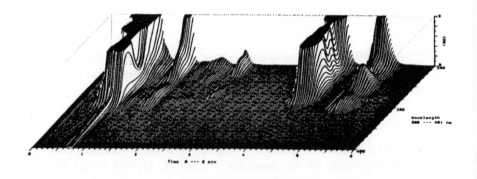

Figure 20: Three dimensional chromatographic plot of a tin plating solution. Compare with Figure 19. From reference 88. Reprinted with permission of the American Electroplaters & Surface Finishers Soc.

ELECTROANALYTICAL TECHNIQUES

Dramatic progress has been made in recent years in electroanalytical methodology, particularly in the areas of polarography, voltammetry and impedance measurements (74,89). Heavy emphasis is increasingly being placed on use of electroanalytical approaches for addressing surface finishing issues, both on the level of fundamental investigations and the control of practical processes. This use of electroanalytical techniques to provide process control of plating solutions is a natural continuation of the methods and technology familiar in any plating shop since the basic principles governing electroplating on a metal substrate and electron transfer at an electrode are the same (89). In each of these situations important items include polarization, current efficiency and iR drop.

Techniques that are being used include dc and ac polarography, differential pulse polarography, square wave voltammetry, linear sweep and cyclic voltammetry, ac cyclic voltammetry and impedance measurements. They will be reviewed briefly in the following pages; for more detail the excellent reviews by Okinaka and Rodgers are recommended (74,89). Table 6 provides examples of use of these techniques for analyzing additives in plating solutions while Table 7 describes the techniques and provides some comparisons. Most of the interest has focused primarily on polarography and voltammetry (89). Both refer to the measurement of a current response to an applied potential. This potential changes with time according to a known function similar to a staircase ramp or pulse.

A. Polarography

A typical instrument for active (analytes in solution are oxidized or reduced by means of an applied potential) electroanalytical techniques is shown in Figure 21 (89). With polarography, a dropping mercury electrode (dme) is used to present a fresh, reproducible surface(Figure 22).

A major improvement to the original dc polarographic method is differential pulse polarography. With this approach, a small amplitude pulse is superimposed on the dc ramp and timed to occur near the end of the lifetime of each mercury drop (Figure 23). With this technique it is possible to obtain enhancement of the detection limit by a factor of at least 100 times as compared to dc polarography (74). To illustrate this sensitivity, differential pulse polarograms of o-chlorobenzaldehyde used as a brightening agent in a chloride zinc solution are shown in Figure 24. This compound has been determined at concentrations as low as 0. 1 ppm (74,90).

Table 7: Comparison of Various Electroanalytical Methods*

Method	Description	Features
DC Polarography	a potential ramp is applied to a dropping mercury electrode and the current is sampled once for each drop	• Most extensively studied; interpretation easiest for fundamental studies • Sensitivity low (10^{-5}M) but not affected by kinetics • Often sufficient for major component analysis • Low resolution for mixture analysis
AC Polarography	superposition of a sinusoidal alternating potential of small amplitude (<10 mV) onto the scanned dc polarization voltage	• Sensitivity higher (10^{-6} to 10^{-7}M) but affected by kinetics • High resolution • Yields information on both faradaic processes and double-layer effects
Differential Pulse Polarography	a small amplitude pulse is super-imposed on the dc ramp and timed to occur near the end of the lifetime of each mercury drop	• Sensitivity high (10^{-7} to 10^{-8}M) but affected by kinetics. • High resolution • Currently most popular for trace analysis
Square-Wave Voltammetry	a single mercury drop is used as the working electrode and the applied potential waveform is a square wave superimposed on the familiar ramp	• Sensitivity even higher than DPP but affected by kinetics • High resolution • Very short analysis time (several seconds) • May dominate in future for trace analysis
Linear Sweep Voltammetry	the applied potential is scanned at a constant rate in one direction only using a solid electrode or a stationary mercury electrode	• Useful for quick diagnostic and mechanistic studies • Fast scan mostly for qualitative work
DC Cyclic Voltammetry	the direction of potential sweep is reversed at a preset value and the current recording continued back to the initial starting potential	• Useful for quick diagnostic and mechanistic studies • Fast scan mostly for qualitative work

(continued)

Table 7: (continued)

Cyclic Voltammetric Stripping	the applied potential is scanned repeatedly between two levels, alternately plating metal from the solution on a solid electrode (reduction) and stripping it off (oxidation).	• For organic additive analysis and mechanistic studies
Cyclic Pulse Voltammetric Stripping	the electrode potential is changed by sequentially stepping between plating, stripping, cleaning, and equilibrating potentials in order to maintain the electrode in a clean and reproducible state	• For organic additive analysis and mechanistic studies
Linear Sweep Stripping Voltammetry	the potential is scanned rapidly in the anodic direction from a suitable initial potential	• For organic additive analysis and mechanistic studies
AC Cyclic Voltammetry	the dc polarization voltage is cycled between two preset values	• Yields information on both faradaic processes and double-layer effects • For studies of organic additive effects and reaction mechanisms
AC Impedance Measurements at Discrete Potentials	charge transfer resistance and double-layer capacitance are determined at discrete dc potentials using one of many available methods for impedance measurement	• For research on reaction mechanisms • For monitoring rate of electroless plating

* **From reference 74.**

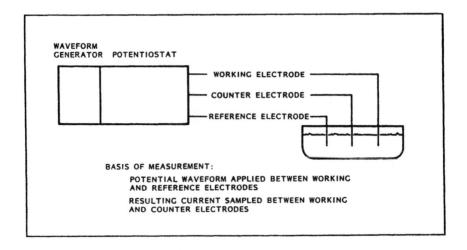

Figure 21: Schematic of typical instrument for active electroanalytical techniques.   From reference 89.   Reprinted with permission of The American Electroplaters & Surface Finishers Soc.

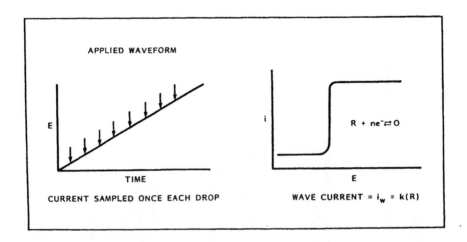

Figure 22: DC polarography.   From reference 89.   Reprinted with permission of The American Electroplaters & Surface Finishers Soc.

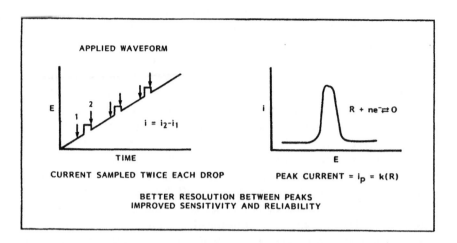

Figure 23: Differential pulse polarography. From reference 89. Reprinted with permission of The American Electroplaters & Surface Finishers Soc.

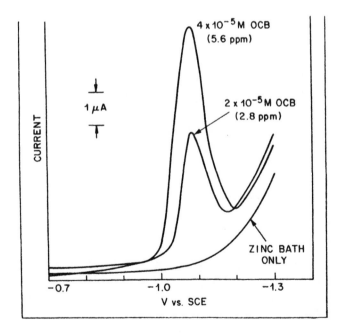

Figure 24: Differential pulse polarograms of o-chlorobenzaldehyde in a chloride zinc plating solution in supporting electrolyte containing 0.6M acetate buffer (pH=5) and 0.04M EDTA. From reference 90. Reprinted with permission of The American Electroplaters & Surface Finishers Soc.

B. Cyclic Voltammetry Stripping

Cyclic voltammetry stripping (CVS) is perhaps the most widely used electroanalytical method for studying electrode processes (74). Tench and Ogden and their circle of researchers have pioneered in the use of CVS, particularly for controlling copper pyrophosphate solutions for printed wiring board production. Their work is detailed in references 91-103 and the following is excerpted from their reports. With CVS, the concentration of brightening or leveling additives in the plating solution is determined from the effect these additives have on the electrodeposition rate (91). The potential of an inert electrode immersed in the solution is cycled as a function of time. As a result of this, a small amount of copper is alternately deposited on the electrode and stripped off by anodic dissolution. A reproducible concentration of brightener is maintained by rotation of the electrode. The technique is illustrated in Figure 25 which shows steady-state voltammetry curves for a Pt disc electrode rotated at 2500 rpm and swept continuously at 50 mV/sec between 0.700 and 1.000 V vs. SCE (saturated calomel electrode) in a pyrophosphate solution containing 1.0 and 2.0 ml/l of proprietary brightener (91). Copper deposition occurs between

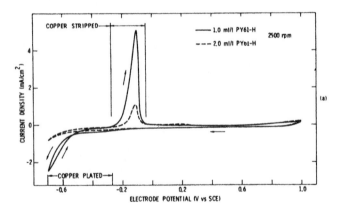

Figure 25: Cyclic voltammograms at 50 mV/sec for a rotating Pt disc electrode in a copper pyrophosphate solution at 22°C. From reference 91. Reprinted with permission of The American Electroplaters & Surface Finishers Soc.

-0.3 and -0.7 V for both sweep directions while the copper deposit is stripped on the anodic sweep between -0.3 and -0.05 V. The area under the stripping peak can be related to the concentration of brightener in the solution since it corresponds to the charge required to oxidize the copper deposit and is proportional to the average deposition rate for that cycle (91). A comparison of the peak areas for solutions containing 1.0 and 2.0 ml/l of proprietary brightener (Figure 25) reveals that the brightener has a strong decelerating effect on the rate of copper deposition. Figure 26 shows that the stripping peak without rotation (Ar) is much larger than that with agitation since the brightener concentration at the electrode surface is nearly zero. This ratio is an effective measure of brightener concentration (91).

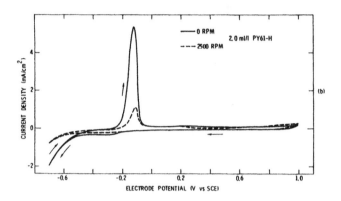

Figure 26: Cyclic voltammograms with and without rotation at 50 mV/sec for a Pt disc electrode in a copper pyrophosphate solution at 22°C. From reference 91. Reprinted with permission of The American Electroplaters & Surface Finishers Soc.

A few examples showing the values of CVS for production control are presented in Figures 27 and 28. One factor that should be taken into account in controlling copper pyrophosphate solutions is the variation of the strength of the additive from batch to batch. Figure 27 shows plots of the CVS rate parameter (Ar/As) for three different batches of additive. Significant differences in both the overall concentration of the active additive and the balance between monomer and dimer are evident. Effective concentration of different batches of additives has been shown to vary by as much as a factor of three (98).

Application of CVS to production copper pyrophosphate solutions reveals considerable information not only about additive behavior but also about the overall functioning of the solution. This is illustrated in Figure 28 which shows voltammograms obtained with the electrode rotation (2500 rpm) in various production solutions (98). Clearly, the CVS method also provides a "fingerprint" that reflects the overall functioning of the solution and in some cases yields specific valuable information.

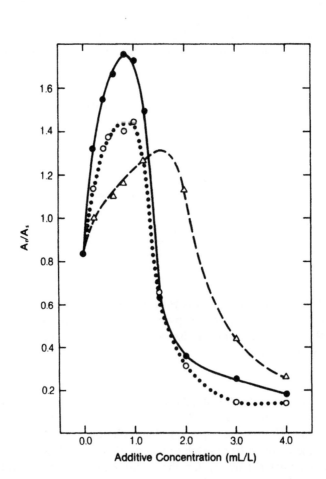

Figure 27: Plots of Ar/As vs. concentration for three batches of additive for a copper pyrophosphate solution. From reference 98. Reprinted with permission of The American Electroplaters & Surface Finishers Soc.

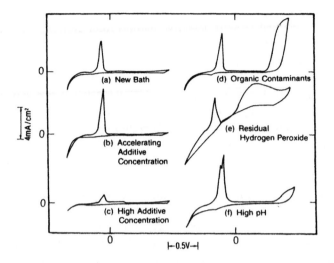

Figure 28: Cyclic voltammograms at 50 mV/sec for a rotating Pt disc electrode (2500 rpm) in various production copper pyrophosphate solutions at 22°C. From reference 98. Reprinted with permission of the American Electroplaters & Surface Finishers Soc.

Solution Condition	Stripping Peak
a. new solution formulated to manufacturers specifications	well defined stripping peak and very little current at more anodic potentials
b. accelerated additive concentrations	stripping peak is much sharper
c. excessive additive levels	stripping peak is strongly suppressed
d. organic contaminants	current wave produced at more anodic potentials reflects solution contamination level
e. residual hydrogen peroxide	this distorted voltammogram reflects residual hydrogen peroxide lingering in solution after treatment with hydrogen peroxide and carbon to remove organic contaminants
f. high solution pH	this produces a splitting of the stripping peak

Another example of the value of CVS for monitoring solution contaminants evolved around defective carbon filter cartridges (98). CVS analysis of a carbon treated solution indicated high amounts of contaminants even when the purification procedure was repeated with a fresh filter cartridge. This led to the discovery that the filter supplier had changed manufacturers and detrimental contaminants were being leached into the solution from the sizing material in the new cartridges. Without CVS to detect this problem, many circuit boards would have been rejected.

Originally developed for pyrophosphate solutions, cyclic voltammetry stripping is now being used for other plating solutions as well (Table 6). However, with acid copper solutions, CVS has two drawbacks: 1) heavily adsorbed films from some acid copper additives undergo successive oxidation/reduction reactions during each analysis cycle and interfere with the copper stripping peak, and 2) most acid copper plating solutions "age" during operation by forming byproducts and this invalidates use of a standard curve of stripping vs. additive concentration generated for a fresh solution (104).

Realizing these difficulties, Tench and White modified the analysis technique to step the electrode potential over a range to include plating, stripping, cleaning and equilibration (103). This technique, called cyclic pulse voltammetric stripping (CPVS), minimizes the effects of contaminants.

C. Impedance Techniques

An electrochemical impedance probe has been developed for monitoring organic plating solution additives (105,106). With this method it is possible to measure increases in interfacial resistance on the surface of a noble metal working electrode and then calibrate in terms of additive concentration. Additives analyzed via this technique are listed in Table 6.

D. Spectrophotometry

Spectrophotometric methods that take advantage of the ultraviolet light absorbance exhibited by many brighteners have been used for determining the concentrations of brighteners in copper (69) and nickel (107).

REFERENCES

1. M. Schwartz, "Deposition From Aqueous Solutions: An Overview", Chapter 10 in *Deposition Technologies for Films and Coatings*, R. F. Bunshah, Editor, Noyes Publications, Park Ridge, New Jersey (1982)

2. D. H. Lashmore, private communication, 1988.

3. U. Landau, "Plating-New Prospects for an Old Art", *Electrochemistry in Industry*, New Directions, U. Landau, E. Yeager and D. Kortan, Editors, Plenum Press, New York (1982)

4. J. Sculley, *Odyssey*, Harper and Row, New York (1987)

5. A. H. Somer, "The Element of Luck in Research-Photocathodes 1930 to 1980", *J. Vac. Sci. Technol.*, Al (2), 119 (Apr-June 1983)

6. R. M. Roberts, *Serendipity*, John Wiley & Sons, New York (1990)

7. A. H. DuRose, "Some Contributions From U.S. Supply Houses to Plating Science and Technology", *Plating* 57, 793 (1970)

8. M. Lyons and W. Millward, British Patent 11,632 (1847)

9. J. A. Henricks, "Bright Electroplating-How We Made the Grade", *Metal Finishing*, 79, 45 (Nov 1981) and 80, 39 (Jan 1982)

10. K. G. Soderberg, "A Glamorous Time-A Look,Back", *Plating* 53, 1262 (1966)

11. T. A. Edison, "Process of Electroplating", U.S. Patent 964,096 (1910)

12. H. Brown, "Electrodeposition of Nickel From an Acid Bath", U.S. Patent 2,191,813 (1940)

13. W. King and E. Todd, U.S. Patent 1,352,328 (1920)

14. C. Humphries, U.S. Patents 1,536,858 and 1,536,859 (1925)

15. H. Geduld, *Zinc Plating*, ASM International, Metals Park, Ohio (1988)

16. J. Pope, private communication, 1981

17.   M. R. Louthan, Jr., "The Effect of Hydrogen on Metals" in *Corrosion Mechanisms*, F. Mansfeld, Editor, Marcel Dekker, New York (1987)

18.   M. Koiwa, A. Yamanaka, M. Arita, and H. Numakura, "Water in Hydrogen Embrittled Copper", Materials Transactions, *J. Inst. Metals*, 30, 991 (1989)

19.   J. R. Denchfield, "Process for Electroforming Low Oxygen Copper", U. S. Patent 3,616,330 (1971)

20.   T. J. Peters and R. H. Waterman, Jr., *In Search Of Excellence*, Warner Books, New York (1984)

21.   *The Properties of Electrodeposited Metals and Alloys, Second Edition*, W. H. Safranek, Editor, American Electroplaters and Surface Finishers Society, Orlando, Florida (1986), First Edition (1974)

22.   D. M. Tench, "Acid Sulfate and Pyrophosphate Copper for Printed Circuits", Short Course Booklet, American Electroplaters and Surface Finishers Society, Orlando, Florida (1982)

23.   C. Struyk and A. E. Carlson, "Nickel Plating From Fluoborate Solutions", *Plating* 37, 1242 (1950)

24.   D. W. Endicott and J. R. Knapp, "Electrodeposition of Nickel-Cobalt Alloy: Operating Variables and Physical Properties of the Deposits", *Plating* 53, 43 (1966)

25.   W. Katz, "Electroforming With Nickel", *Metall* (Berlin), 21, 580 (1967)

26.   N. P. Fedot'ov, "Physical and Mechanical Properties of Electrodeposited Metals", *Plating* 53, 309 (1966)

27.   A. Brenner and C. W. Jennings, *Proceedings American Electroplaters Soc.*, 35, 31 (1948)

28.   C. Ogden, D. Tench and J. White, "Tensile Properties and Morphology of Pyrophosphate Copper Deposits", *Journal of Applied Electrochemistry*, 12, 619 (1982)

29.   G. T. Rogers and K. J. Taylor, "Advances in Electroplating Research", *Electrical Review*, 125 (Jan 22, 1971)

30.   S. E. Beacom and B. J. Riley, "Tracer Follows Leveler in Electroplating Bath", *Nucleonics*, 18, No 5, 82 (May 1960)

31.   J. Edwards and M. Levitt, "Radiotracer Study of Addition Agent Behavior", *Trans. Inst. Metal Finishing*, Parts I, II, and III, 39, 33 (1962), Parts IV, V, and VI, 41, 140, 147 and 157 (1964), Part VII, 44, 27 (1967) and Part VIII, 45, 12 (1967)

32.   S. E. Beacom and B. J. Riley, "A Radioisotope Study of Leveling in Bright Nickel Electroplating Baths", *J. Electrochem. Soc.*, 106, 309 (1959), "Further Studies of Leveling Using Radiotracer Techniques", 107, 785 (1960), and "Mechanism of Addition Agent Reaction in Bright and Leveling Nickel Deposition, 1. Studies With Radioactive Sodium Allyl Sulfonate",108, 758 (1961)

33.   G. T. Rogers, M. J. Ware and R. V. Fellows, "The Incorporation of Sulfur in Electrodeposited Nickel Using Thiourea as a Brightener and Leveler", *J. Electrochem. Soc.*, 107, 677 (1960)

34.   J. Hoekstra and D. Trivich, "The Uptake of Sulfur From Plating Brighteners by Copper and Nickel", *J. Electrochem. Soc.*, 111, 162 (1964)

35.   G. T. Rogers and K. J. Taylor, "The Reactions of Coumarin, Cinnamyl Alcohol, Butynediol, and Propargyl Alcohol at an Electrode on Which Nickel is Depositing", *Electrochim. Acta*, 11, 1685 (1966)

36.   W. R. Doty and B. J. Riley, "A Study of the Inclusion of a Radioactive Addition Agent in Bright Nickel", *J. Electrochem. Soc.*, 114, 50 (1967)

37.   W. H. Safranek and J. G. Beach, "Electroplating Bright Leveling and Ductile Copper", CDA Report 122/8, Copper Development Association, Greenwich, CT (March 1968)

38.   F. A. Lowenheim, *Electroplating*, McGraw-Hill, New York (1978)

39.   R. Weil and R. Paquin, "The Relationship Between Brightness and Structure in Electroplated Nickel", *J. Electrochem. Soc.*, 107, 87 (1960)

40.   R. Weil, "Structure, Brightness and Corrosion Resistance of Electrodeposits", *Plating*, 61, 654 (1974)

41.   E. H. Lyons, Jr., "Fundamental Principles", in *Modern Electroplating*, Third Edition, F. A. Lowenheim, Editor, John Wiley and Sons, New York (1974)

42.   D. A. Vermilyea, "Additives and Grain Refinement", *J. Electrochem. Soc.*, 106, 66 (1959)

43.   C. C. Roth and H. Leidheiser, Jr., "The Interaction of Organic Compounds With the Surface During the Electrodeposition of Nickel", *J. Electrochem. Soc.*, 100, 553 (1953)

44.   S. Nakahara and R. Weil, "Initial Stages of Electromonocrystallization of Nickel on Copper-Film Substrates", *J. Electrochem. Soc.*, 120, 1462 (1973)

45.   H. Brown, "Addition Agents, Anions, and Inclusions in Bright Nickel Plating", *Plating* 55, 1047 (1968)

46.   J. K. Dennis, "Decorative Electrodeposited Metallic Coatings", *International Metallurgical Reviews*, 17, 43 (1972)

47.   J. Reid, "An HPLC Study of Acid Copper Brightener Properties", PC Fab 10, 65 (Nov 1987)

48.   F. Passal, "Copper Plating During the Last Fifty Years", *Plating*, 46, 628 (1959)

49.   D. G. Foulke, "Acid Electrolytes", Chapter 6 in *Gold Plating Technology*, F. H. Reid and W. Goldie, Electrochemical Publications Ltd. (1974)

50.   J. D. E. McIntyre and W. F. Peck, Jr., "Electrodeposition of Gold: Depolarization Effects Induced by Heavy Metal Ions", *J. Electrochem. Soc.*, 123, 1800 (1975)

51.   J. D. E. McIntyre, S. Nakahara and W. F. Peck, Jr., "Structural Characteristics of Soft Gold Electrolytes", *J. Electrochem. Soc.*, 124, 302C (1977)

52.   R. J. Morrissey and A. M. Weisberg, "Some Further Studies on Porosity in Gold Electrodeposits", in Corrison Control by Coatings, H. Leidheiser, Jr., Editor, *Science Press* (1979)

53.   G. Bacquias, "Bright Gold Electroplating Studies", *Gold Bulletin*, 15 (4), 124 (1982)

54.   J. W. Dini and J. R. Helms, "Properties of Electroformed Lead", *Metal Finishing*, 67, 53 (August 1969)

55.   A. K. Graham and H. L. Pinkerton, "Evaluation of Addition Agents for High-Speed Lead Plating", 50th Annual Technical Proceedings, *American Electroplaters Soc.*, 139 (1963)

56.   E. B. Saubestre, "The Chemistry of Bright Nickel Plating Solutions", *Plating* 45, 1219 (1958)

57.   G. Dubpernell, "The Story of Nickel Plating", *Plating* 46, 599 (1959)

58.   H. Brown and B. B. Knapp, "Nickel Plating", Chapter 12 in *Modern Electroplating*, Third Edition, F. A. Lowenheim, Editor, John Wiley & Sons, New York (1974)

59.   M. J. Levett, A. F. Cudd and J. L. Cox, "A Note on the Effects of Addition Agents on the Microstructure of Electrodeposited Nickel Plate", *J. Institute of Metals*, 96, 125 (1968)

60.   B. M. Luce and D. G. Foulke, "Silver", Chapter 14 in *Modern Electroplating*, Third Edition, F. A. Lowenheim, Editor, John Wiley & Sons, New York (1974)

61.   A. H. DuRose, *Modern Electroplating*, 2nd Edition, F. A. Lowenheim, Editor, John Wiley & Sons, New York, p 377 (1963)

62.   R. M. MacIntosh, "Acid Tin Plating", Chapter 15 in *Modern Electroplating*, Third Edition, F. A. Lowenheim, Editor, John Wiley & Sons, New York (1974)

63.   P. A. Kohl, "The High Speed Electrodeposition of Sn/Pb Alloys", *J. Electrochem. Soc.*, 129, 1196 (1982)

64.   A. K. Graham and H. L. Pinkerton, "Properties and Comparative Performance of Electrodeposited Terne Alloys", *Plating* 54, 367 (1967)

65.   S. A. Tucker and E. G. Thomssen, "Electrolytic Deposition of Lead and Zinc as Affected by the Addition of Certain Organic Compounds", *Trans. Amer. Electrochem. Soc.*, XV, 477 (1909)

66.   M. R. Thompson, "Acid Zinc Plating", *Trans. Amer. Electrochem. Soc. L.* 193 (1926)

67.   E. C. Broadwell, U. S. Patent, 905,837 (Dec 1908)

68.   G. Langbein and W. T. Brannt, *Electrodeposition of Metals*, Eighth Edition, Henry Casey Baird & Co., New York, p. 556 (1920)

69.   D. E. Crotty, "The Fundamentals of Zinc Brightener Design", Proceedings SUR/FIN 88, *American Electroplaters & Surface Finishers Soc.* (1988)

70.   B. DaFonte, Jr., "Recent Developments in Non-Cyanide Zinc Plating", Proceedings SUR/FIN 78, *American Electroplaters & Surface Finishers Soc.* (1978)

71.   D. M. Hembree, Jr., "Linear Sweep Voltammetry to Determine Lignin Sulfonate in Lead Plating Baths", *Plating & Surface Finishing*, 73, 54 (Nov 1986)

72.   T. C. Franklin, "Some Mechanisms of Action of Additives in Electrodeposition Processes", *Surface and Coatings Technology*, 30, 415 (1987)

72a.   L. Oniciu and L. Muresan, "Some Fundamental Aspects of Levelling and Brightening in Metal Electrodeposition", *Journal of Applied Electrochemistry* 21, 565 (1991)

73.   C. J. Bocker and Th. Bolch, "Nickel Electroforming-Some Aspects For Process Control", Proceedings AES International Symposium on Electroforming/Deposition Forming, Los Angeles, CA. (March 1983)

74.   Y. Okinaka, "Electroanalytical Methods for Research and Development in Electroplating", *Plating & Surface Finishing*, 72, 34 (Oct 1985)

75.   J. L. Jackson and D. A. Swalheim, "Hull Cell Tests for Quality Plating", Illustrated Slide Lecture Text, *American Electroplaters & Surface Finishers Soc.*, Orlando, FL (1972)

76.   J. L. Jackson and D. A. Swalheim, "The Hull Cell-An Indispensable Tool", *Plating & Surface Finishing*, 65, 38 (May 1978)

77.   R. O. Hull, U. S. Patent 2,149,344 (1939)

78.   M. K. Sanicky, "All About the Hull Cell", *Plating & Surface Finishing*, 72, 20 (Oct 1985)

79.   Anon., "Finishing Pointer-Graphic Description of a Hull Cell Panel", *Metal Finishing*, 75, 34 (July 1977)

80.   J. P. Branciaroli, "Modified Hull Cell", *Plating* 46, 257 (1959)

81.   "The Gornall Cell", Technical Data Sheet, McGean-Rohco, Inc., Cleveland, Ohio (1986)

82.   W. Nohse, *The Hull Cell*, Robert Draper Ltd., Teddington, England (1966)

83.   M. L. Rothstein, "Control of Plating Chemistry by Voltammetric Techniques", *Plating & Surface Finishing*, 71, 36 (Nov 1984)

84.   A. E. Olsen, "Bright Acid Sulfate Copper Plating", Illustrated Slide Lecture Text, American Electroplaters & Surface Finishers Soc., Orlando, FL (1970)

85.   P. Jandik, W. R. Jones, B. J. Wildman, J. Krol and A. L. Heckenberg, "Liquid Chromatographic Methods for Monitoring of Plating Baths Used in Manufacturing of PC Boards", *Proceedings SURIFIN 89*, American Electroplaters & Surface Finishers Soc., Orlando, FL (1989)

86.   L. R. Snyder and J. J. Kirkland, *Introduction to Modern Liquid Chromatography, 2nd Edition*, John Wiley & Sons, New York (1979)

87.   S. S. Heberling, D. Campbell and S. Carson, "Monitoring Acid Copper Plating Baths", *PC Fab*, 12, 72 (August 1989)

88.   P. Jandik, A. Heckenberg, R. L. Lancione, and B. Kahler, "Applications of a Photodiode Array UV Detector to Chromatographic Analysis of Organic Additives in Plating Solutions", *Proceedings AESF Analytical Methods Symposium*, American Electroplaters & Surface Finishers Soc., Orlando, FL (Jan 1988)

89.   T. Rodgers, "Electroanalytical Methods in the Plating Industry", An Overview", *Plating & Surface Finishing*, 74, 51 (Nov 1987)

90.   H. F. Bell, "Applications of Pulse Polarography", *Proceedings Fourth Plating in the Electronics Industry Symposium*, American Electroplaters & Surface Finishers Soc., Orlando, FL (1973)

91.   C. Ogden and D. Tench, "Optimized Circuit Board Plating With Copper Pyrophosphate Baths", *Plating & Surface Finishing*, 66, 30 (Sept 1979)

92.   D. Tench and C. Ogden, "A New Voltammetric Stripping Method Applied to the Determination of the Brightener Concentration in Copper Pyrophosphate Plating Baths", *J. Electrochem. Soc.*, 125, 194 (1978)

93.   D. Tench and C. Ogden, "On the Functioning and Malfunctioning of Dimercaptothiadiazoles as Leveling Agents in Circuit Board Plating for Copper Pyrophosphate Baths", *J. Electrochem. Soc.*, 125, 1218 (1978)

94.   C. Ogden and D. Tench, "Control of Contaminants in Copper Pyrophosphate Circuit Board Production Baths", *Plating & Surface Finishing*, 66, 45 (Dec 1979)

95.   C. Ogden and D. Tench, "On the Mechanism of Electrodeposition in the Dimercaptothiadiazole/Copper Pyrophosphate System", *J. Electrochem. Soc.*, 128, 539 (1981)

96.   R. Haak, C. Ogden and D. Tench, "Cyclic Voltammetric Stripping Analysis of Acid Copper Sulfate Plating Baths, Part 1-Polyether-Sulfide-Based Additives", *Plating & Surface Finishing*, 68, 52 (April 1981)

97.   R. Haak, C. Ogden, and D. Tench, "Cyclic Voltammetric Stripping Analysis of Acid Copper Sulfate Plating Baths-Part 2, Sulfoniumalkanesulfonate-Based Additives", *Plating & Surface Finishing*, 69, 62 (March 1982)

98.   B. Lowry, C. Ogden, D. Tench and R. Young, "Production Implementation of Controls for Copper Pyrophosphate Circuit Board Plating Baths", *Plating & Surface Finishing* 70, 70 (Sept 1983)

99.   W. O. Freitag, C. Ogden, D. Tench and J. White, "Determination of the Individual Additive Components in Acid Copper Plating Baths", *Plating & Surface Finishing* 70, 55 (Oct 1983)

100.   D. Anderson, M. Hanna, C. Ogden and D. Tench, "Purification of Acid Copper Plating Baths", *Plating & Surface Finishing* 70, 70 (Dec 1983)

101.   M. Jawitz, C. Ogden, D. Tench and R. Thompson, "Optimization of Copper Pyrophosphate Circuit Board Plating Baths at High Additive Concentrations", *Plating & Surface Finishing*, 71, 58 (Jan 1984)

102.   C. Ogden, and D. Tench, "Cyclic Voltammetric Stripping Analysis of Copper Plating Baths", Application of Polarization Measurements in the Control of Metal Deposition, I. H. Warren, Editor, Elsevier (1984)

103.   D. Tench and J. White, "Cyclic Pulse Voltammetric Stripping Analysis of Acid Copper Plating Baths", *J. Electrochem. Soc.*, 132, 831 (1985)

104.   G. L. Fisher and P. J. Pellegrino, "The Use of Cyclic Pulse Voltammetric Stripping for Acid Copper Plating Bath Analysis", *Plating & Surface Finishing*, 75, 88 (June 1988)

105.   J. C. Farmer and W. D. Bonivert, "Impedance Probe for Measuring Organic Additives in Electroplating Baths", *Plating & Surface Finishing* 76, 56 (Aug 1989)

106.   J. C. Farmer and H. R. Johnson, "Application of Dynamic Impedance Measurements for Adsorbed Plating Additives", *Plating & Surface Finishing*, 72, 60 (Sept 1985)

107.   W. G. Ward, "Analysis of Nickel Plating Solutions", Proceedings SUR/FIN 89, *American Electroplaters & Surface Finishers Soc.*, Orlando, FL (1989)

108.   G. L. Fisher, "CPVS Analysis of Acid Copper Plating Baths Used for Printed Wiring Boards", *Proceedings SUR/FIN 89*, American Electroplaters & Surface Finishers Soc., Orlando, FL (1989)

109.   W. O. Freitag and M. R. Manning, "Analysis of Additives in Acid Copper Baths by Cyclic Voltammetry", *Proceedings SUR/FIN 89*, American Electroplaters & Surface Finishers Soc., Orlando, FL (1989)

110.   D. Engelhaupt and J. Canright, "Additive Analysis in PWB Copper and Solder Plating Processes", *Proceedings SUR/FIN 84*, American Electroplaters & Surface Finishers Soc., Orlando, FL (1984)

111.   J. Van Puymbroeck and J. VanHumbeeck, "Voltammetric Method for the Control of the Activity of Additives in Acid Copper Electrolytes", *Proceedings 2nd AESF Analytical Methods Symposium*, American Electroplaters & Surface Finishers Soc., Orlando, FL (Jan 1988)

112.   R. Gluzman, "Control of Production Copper Electroplating Baths Using Cyclic Voltammetric Stripping Analysis", *Trans. Inst. Metal Finishing*, 63, 134 (1985)

113.   S. Heberling, D. Campbell and S. Carson, "On-Line Measurement and Control of Organic Additives in Acid Copper Baths Using Liquid Chromatography", Proceedings SUR/FIN 89, American Electroplaters & Surface Finishers Soc., Orlando, FL (1989)

114.   A. Gemmler and T. Bolch, "Analysis of Wetting Agents and Brighteners in Acid Copper Plating Baths:New Aspects for Process Control", Proceedings SUR/FIN 87, American Electroplaters & Surface Finishers Soc., Orlando, FL (1987)

115.   D. A. Zatko and W. O. Freitag, "Analysis of Organic Components in Copper Plating Baths by Liquid Chromatography", Proceedings SUR/FIN 83, American Electroplaters & Surface Finishers Soc., Orlando, FL (1983)

116.   M. Grall, G. Durand and J. M. Couret, "Cyclic Voltammetric Determinations of Brightener Concentration in Acid Copper Plating Baths", Plating & Surface Finishing, 72, 72 (Dec 1985)

117.   S. A. Edwards and J. Reid, "Chromatographic Studies and Monitoring of Acid Copper Brighteners", Proceedings SUR/FIN 89, American Electroplaters & Surface Finishers Soc., Orlando, FL (1989)

118.   M. L. Rothstein, W. M. Peterson and H. Siegerman, "Polarographic Methods of Analyzing Plating Baths", Plating & Surface Finishing, 68, 78 (June 1981)

119.   J. C. Farmer, "Underpotential Deposition of Copper on Gold and the Effects of Thiourea Studied by AC Impedance", J. Electrochem. Soc., 132, 2640 (1985)

120.   J. D. Reid and A. P. David, "Effects of Polyethylene Glycol on the Electrochemical Characteristics of Copper Cathodes in an Acid Copper Medium", Plating & Surface Finishing, 74, 66 (Jan 1987)

121.   G. Bush, "The Use of Cyclic Voltammetric Stripping for Determination of Additive Concentration in Copper Plating Baths", Proceedings SUR/FIN 83, American Electroplaters & Surface Finishers Soc., Orlando, FL (1983)

122.   M. Paunovic and R. Arndt, "The Effect of Some Additives on Electroless Copper Deposition", J. Electrochem. Soc., 130, 794 (1983)

123.   L. E. Fosdick, "Electroless Plating Bath Analysis Using Voltammetric Techniques", Proceedings Second AES Electroless Plating Symposium, American Electroplaters & Surface Finishers Soc., Orlando, FL (Feb 1984)

124.   R. Haak and D. Tench, "Polarographic Analysis of Hard Gold Plating Baths, *Plating & Surface Finishing* 73, 72 (June 1986)

125.   J. C. Farmer and R. H. Muller, "Effect of Rhodamine-B on the Electrodeposition of Lead on Copper", *J. Electrochem. Soc.*, 132, 313 (1985)

126.   M. Carano and W. Ward, "Determination of Pyridine Based Leveling Compounds in Nickel Electroplating Baths by Differential Pulse Polarography", *Plating & Surface Finishing*, 71, 54 (Nov 1984)

127.   A. Palus, "The Determination of Plating Bath Additives Using HPLC and Electrochemical Techniques", *Proceedings SUR/FIN 89*, American Electroplaters & Surface Finishers Soc., Orlando, FL (1989)

128.   K. Haak, "Ion Chromatography in the Electroplating Industry", *Plating & Surface Finishing* 70, 34 (Sept 1983)

129.   J. J. Reiss, L. Ashley and S. K. Bohra, "Analysis of Plating Baths by Differential Pulse Polarography", *Proceedings SUR/FIN 83*, American Electroplaters & Surface Finishers Soc., Orlando, FL (1983)

130.   D. Crotty and K. Bagnall, "Analysis Methods for Zinc Plating Brighteners", *Plating & Surface Finishing*, 75, 52 (Nov 1988)

131.   F. Chassaing, M. Joussellin and R. Wiart, "The Kinetics of Nickel Electrodeposition Inhibition by Adsorbed Hydrogen and Anions", *J. Electroanal. Chem.*, 157, 75 (1983)

132.   J. P. G. Farr and O. A. Ashirv, "Electrode Impedance and the Effect of Additives on the Electrodeposition of Silver", *Trans. Inst. Metal Finishing*, 64, 137 (1986)

133.   A. L. Johnson, S. Morse, E. Winterrowd and W. G. Ward, "Reverse Phase Liquid Chromatography-Determination of Additives in Bright Acid Tin Plating Solutions", *Metal Finishing*, 83, 49 (Oct 1985)

134.   W. G. Ward, A. L. Johnson, S. Morse and E. Winterrowd, "HPLC Analysis of Additives in Bright Acid Tin Plating Solutions", *Proceedings 2nd AESF Analytical Methods Symposium*, American Electroplaters & Surface Finishers Soc., Orlando, FL (Jan 1988)

135.   P. Bratin, M. Kerman and R. Gluzman, "Recent Developments in CVS Analysis of Plating Solutions", *Proceedings SUR/FIN 87*, American Electroplaters & Surface Finishers Soc., Orlando, FL (1987)

136.   R. L. Summers, "Differential Pulse Polarography of Organic Additives in Acid Tin and Bright Tin-Lead Plating Baths", *Proceedings SUR/FIN 83*, American Electroplaters & Surface Finishers Soc., Orlando, FL (1983)

8
POROSITY

INTRODUCTION

Voids (pores) are generally formed in thin films irrespective of the film preparation method (electrodeposition, evaporation, or sputtering) as long as the deposition process involves a phase transformation from the vapor to the solid state. These voids can be extremely small (approximately 10Å) and high in density (about $1 \times 10^{17}/cm^3$)(1-3).

Porosity is one of the main sources of discontinuities in electroplated coatings; the others are cracks from high internal stresses and discontinuities caused by corrosion or subsequent treatments such as wear of deposits after plating as shown in Figure 1 (4,5). In most cases porosity is undesirable. Pores can expose substrates to corrosive agents, reduce mechanical properties, and deleteriously influence density, electrical properties and diffusion characteristics. As discussed in the chapter on diffusion, pores formed as a result of heating (Kirkendall voids) can noticeably reduce adhesion of a deposit.

Porosity in a sacrificial coating such as zinc on steel is not too serious since in most environments zinc will cathodically protect steel at the bottom of an adjacent pore. However, for a noble metal, similar porosity may be problematic. A special significance of porosity is that it permits the formation of tarnish films and corrosion products on the surface, even at room temperature. In the electronics industry, which utilizes the largest quantity of gold coatings for engineering purposes, porosity is a major concern because of its effect on the electrical properties of plated parts (6). Porosity in cadmium deposits, which is desirable for purging hydrogen codeposited during plating, can result in rapid postplating embrittlement due

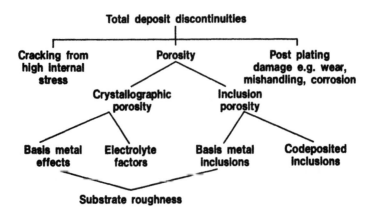

Figure 1:   Causes of discontinuities in electroplated coatings. Adapted from Reference 4.

to the lack of a barrier to hydrogen reentering the steel during exposure of the plated part to corrosive environments (7,8).

Depending on the method of formation of the coating, the pore can be filled with air or foreign matter such as gases, fluids, solids, etc. For example, analysis of electrodeposits reveals small amounts of many constituents from plating solutions easily explained by solution filled cavities of small pores but difficult to account for otherwise. Outgassing measurements on electrodeposited gold films revealed that the major constituent in voids in these coatings was hydrogen gas (4,9).

INFLUENCE ON PROPERTIES

Any material (coatings, castings, powder metallurgy consolidated alloys, etc.) containing pre-existing porosity or voids is subject to property degradation. The tensile behavior of materials with pre-existing porosity is characterized by large decreases in both strength and ductility with increasing porosity level, since ductile fracture in engineering alloys is most often the result of the nucleation and link-up of voids or cavities (10). Figure 2 reveals that both powder metallurgy Ti-6Al-4V and chemically pure Ti suffer a decrease in yield strength as well as tensile ductility with increasing porosity level. Three percent porosity in cast, high-purity copper, which reduces the density from 8.93 to 8.66 gm/cm^3 drops the reduction-in-area at 950°C from 100% to 12% (11). Porosity introduces two factors which reduce macroscopic ductility. First, the presence of pores acts to concentrate strain in their vicinity and to reduce the macroscopic flow stress. Secondly, the nonregular distribution of pores results in paths

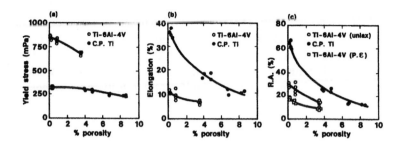

Figure 2:   The influence of porosity on (a) the yield stress, (b) the elongation to failure, and (c) the percent reduction in area for chemically pure titanium and Ti-6Al-4V. Adapted from Reference 10.

of high pore content which are preferred sites for flow localization and fracture (10).

Table 1 shows the influence of porosity on various mechanical and physical properties of thin films (1). Point defects such as pinholes laid down during deposition and generated during thermal cycling may act as starting points for severe film cracking at high temperature. Tests carried out on evaporated coatings of chromium, copper and nickel showed that cracks radiated from pinholes in the films. This effect was attributed to stress concentration in the neighborhood of the pinhole (12). Pre-existing

Table 1:   Effects of Voids on the Properties of Thin Films*

Properties	Effects of Voids
Mechanical properties	Ductility decrease
	Hydrogen embrittlement
	Creep resistance
	Reduced elastic modulus
	Decrease in adhesion(interfacial void)
Electrical properties	Resistivity increase
Corrosion properties	Reduced corrosion resistance (through-pores)
Dielectric properties	Dielectric constant

*From reference 3.

voids, along with hydrogen, are responsible for the reduced ductility of electroless copper deposits (13). This is discussed in more detail in the chapter on hydrogen embrittlement. Chromate coatings on copper and nickel-phosphorus films prepared by electrodeposition also contain a high density of voids with a structure similar to that of a crack network. The presence of these voids contributes significantly to brittleness in these films (14,15).

GOOD ASPECTS ABOUT POROSITY

There are occasions where porosity is desired in a coating. Pores in anodized aluminum provide the opportunity to provide a wide range of colors when they are sealed to eliminate the path between the aluminum and the environment, and pores in phosphoric acid anodized aluminum provide for adhesion of subsequent deposits. Porous chromium deposits from specially formulated solutions provide for improved lubricating properties while microporous chromium deposits, produced by plating over a nickel deposit which contains codeposited multitudinous fine, nonconducting particles, result in uniform distribution of corrosion attack of the nickel (16). Porous electroforms for applications such as perforated shells used in vacuum forming procedures or for fluid retention have been produced (17-19). One technique involved addition of graphite particles to a nickel plating solution. The graphite particles adhered to the deposit and generated channels 50-100/μm in diameter which were propagated through the nickel for 2.5mm or more (17). Another approach involved codeposition of nonconducting powders with the nickel and by decomposing the powders at a low temperature after plating, horizontal as well as vertical porosity was achieved (18).

CLASSIFICATION OF PORES

Kutzelnigg suggests that pores may be broken down into two main categories, transverse pores and masked or bridged pores (20). His pictorial descriptions of the various types of cavities are shown in Figure 3 and the following information is extracted from his comprehensive article on porosity (20). Transverse pores may be either of the channel type (Figure 3a) or hemispherical (Figure 3b) and extend through the coating from the basis metal to the surface of the deposit. They may be oriented perpendicular (Figures 3a,b) or oblique (Figure 3c) to the surface or may have a tortuous shape (Figure 3d). Masked or bridged pores do not extend through the coating to reach the surface but either start at the surface of the basis metal and become bridged (Figure 3e) or start within the coating and become bridged (enclosed pores) (Figure 3f). A pit is a surface pore which does not become masked or bridged (Figure 3g). They may be hybrids (Figure 3h), or give rise to blisters (Figure 3i). Cracks may be regarded as pores much extended in a direction parallel to the surface, but they can also be divided into transverse cracks, enclosed cracks and surface cracks (Figures 3j, 3k, 3l).

A combination of channel and spherical pores is shown in Figure 3m and the influence of substrate defects in Figures 3n, 3o, and 3p. Chemical attack after deposition (Figure 3q), incomplete coverage of the deposit (Figure 3r), and defects due to inclusions (Figures 3s and 3t) are other examples of pores (20).

CAUSES OF POROSITY

Porosity, together with structure and many other properties of an electroplated coating, reflects the effects of: 1) nature, composition and history of the substrate surface prior to plating; 2) composition of the plating solution and its manner of use; and 3) post plating treatments such as polishing (abrasive or electrochemical) wear, deformation, heating and corrosion (21).

A pore may arise in several ways: 1) irregularities in the basis metal; 2) local screening of the surface to be coated; 3) faulty conditions of deposition; and 4) damage after plating. The first two may be attributed to inadequacies of prior processing such as cleaning, pickling, rolling, machining, heat treating, etc. (20). Number three is related to the ability of the plating process to adequately cover the surface through the conventional steps of nucleation and growth. If lateral growth can be promoted in place of outward growth of the deposit, coverage is faster and therefore more effective at lower thickness (22,23,24), as will be shown later in this chapter.

Figure 1 shows that porosity is caused by either inclusions (inclusion porosity) or by misfit of crystal grains (crystallographic porosity). Inclusion porosity arises from small nonconducting areas on the substrate which are not bridged over during the early stages of deposition. Crystallographic porosity arises from structural defects caused by either the basis metal or electrolyte factors (4).

At low deposit thickness, porosity of electrodeposited films is largely controlled by the surface condition and characteristics of the underlying substrate. This condition persists up to a limiting thickness, after which the properties of the film itself, primarily crystallographic properties, determine the rate of pore closure (22). Typically, porosity drops

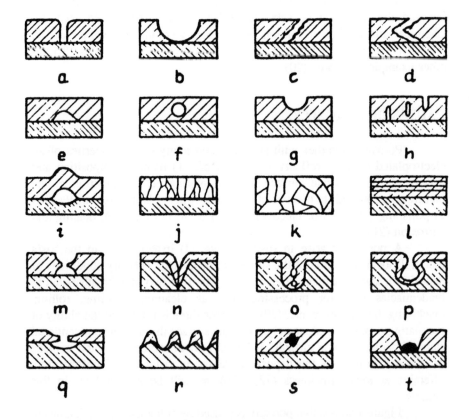

Figure 3: Types of pores or cavities. From Reference 20. Reprinted with permission of the American Electroplaters & Surface Finishers Society.

a   Transverse pore oriented perpendicular and extending through the coating from the basis metal channel pore.

b   Same as a) but this pore is hemispherical.

c   Transverse pore extending through the coating in an oblique fashion.

d   Transverse pore extending through the coating in a tortuous fashion.

e   Masked or bridged pore-starts at the surface of the basis metal but does not reach the surface of the deposit.

f   Masked or bridged pore-starts within the deposit and becomes bridged (enclosed pore).

g   A pit-which does not reach the surface of the basis metal (dead end pore).

h   A hybrid-a bridged pore in contact with the base, an enclosed pore, and a surface pit.

i   Bridged pores located on the surface of the base metal and originally filled with electrolyte may give rise to "blisters" if the deposit is locally lifted by the pressure of hydrogen generated by interaction of the basis metal and the solution. Blisters may also be produced by rubbing poorly adherent deposits (or heating them).

j   Cracks-may be regarded as pores much extended in a direction parallel to the surface. Cracks may also be divided into transverse cracks, enclosed cracks, and surface cracks. They may further be gross, small, or submicroscopic. An extreme case of the last type are the boundaries of the crystallites building up in the deposit.

k   The most common examples of cracks is represented by the pattern seen in bright chromium deposits at large magnification.

l   Stratifications which may be better understood as lamellar discontinuties. In general these discontinuities differ in composition from the main part of the deposit.

m   A combination of channel type and spherical pores.

n   Example of porosity obtained with a V notched substrate.

o   Example of porosity obtained with a U notched substrate.

p   Another type of trouble may arise from pores in the basis metal, e.g., a casting or powder compact part. Though the deposit itself may be free of pores, the resulting pocket filled with electrolyte is the cause of trouble known as blooming out.

q   Chemical attack after deposition.

r   Incomplete coverage of the surface due to poor macro- or micro-throwing power of the solution (also applies to n and o).

s   A defect due to an inclusion-finely dispersed oxide, hydroxide, sulfide, basic matter or as adsorbed organic compounds.

t   Another defect due to an inclusion-carbon particles from overpickled steel, residues of polishing compounds, etc.

exponentially with thickness as shown in Figure 4 (23).

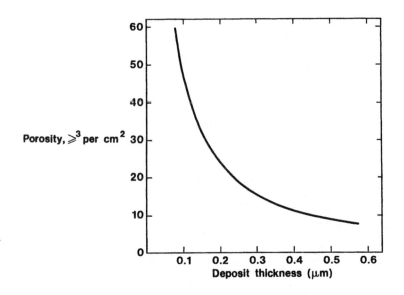

Figure 4: Variation of coating porosity with thickness for electrodeposited chromium. Adapted from reference 23.

An example of the sensitivity of porosity to substrate and deposition parameters is illustrated in Figure 5 which shows three distinct phases for electrodeposited, unbrightened gold on a copper substrate: substrate dominated, transition, and coating dominated. For very thin gold coatings (less than about 1µm), substrate texture controls coating porosity. At greater thicknesses, the slope of the porosity-thickness curve is controlled by parameters relevant to the deposit itself. Between these two regimes is a sharp, well marked transition region in which the porosity of the deposit falls extremely rapidly. The thickness at which this sharp transition occurs varies with the deposit grain size.   The form and position of the porosity-thickness plots are affected by the deposit grain size, the crystallographic orientation and the ratio of nucleation rate to rate of grain growth, which, in turn, controls the average grain size of the deposit at any given thickness (22,24,25).

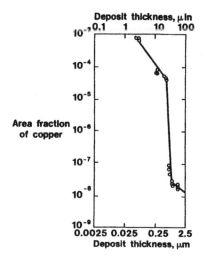

Figure 5:   Porosity versus deposit thickness for electrodeposited unbrightened gold on a copper substrate. Adapted from Reference 22.

FACTORS RELATING TO THE SUBSTRATE

The surface of a substrate has small areas with the property of initiating pores which are referred to as pore precursors (26). These precursors prevent fusion of crystals and as the coating thickens a pore is generated. Inclusions of slag, oxides, sulfides, polishing abrasive, dirt, subscale oxide, and particles settling on the substrate from the plating solution, are pore precursors (21).

Substrate surface roughness has a noticeable influence on porosity. This is shown in Figures 6 and 7 for pure acid citrate gold plated directly on OFHC copper discs (Figure 6) and OFHC copper discs with a nickel underplate (Figure 7). The data clearly show a large increase in porosity with roughness. Rough surfaces have a true area greater than the apparent area (Figure 8). Therefore, it is quite possible that at least some of the increase in porosity on rough substrates compared to the porosity on smooth substrates is due to a difference in average true plate thickness. Garte proposed a roughness factor ratio to help explain this (27):

$$R \text{ (roughness factor)} = \frac{\text{True Area}}{\text{Apparent Area}}$$

If the plating thickness, T, is determined by a weight per unit (apparent) area method, as most methods are, then

$$T = \frac{\text{Weight}}{\text{Apparent   Area}} \times \frac{1}{\text{Density}}$$

$$\text{Therefore: True   Thickness(t)} = \frac{\text{Measured   Thickness (T)}}{R}$$

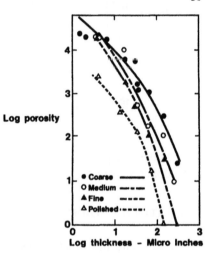

Figure 6: Relationship between porosity-thickness-roughness for acid citrate gold on OFHC copper. Adapted from Reference 27.

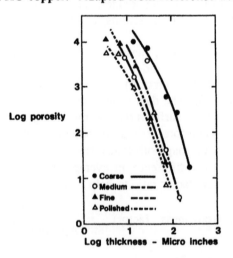

Figure 7: Relationship between porosity-thickness-roughness for acid citrate gold on OFHC copper with a nickel underplate. Adapted from Reference 27.

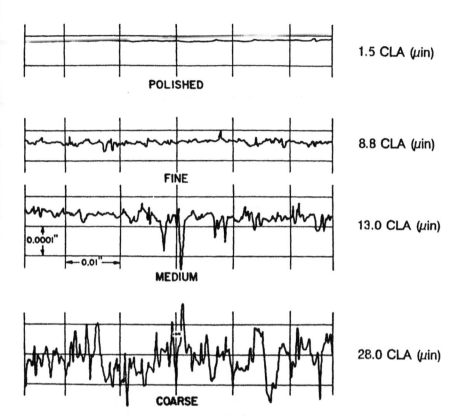

1.5 CLA (µin)

POLISHED

8.8 CLA (µin)

FINE

13.0 CLA (µin)

0.0001"

←— 0.01"—→

MEDIUM

28.0 CLA (µin)

COARSE

Figure 8: Diamond stylus profilometer tracings of specimen surfaces. Original magnification: 5000× vertically, 100× horizontally. Reprinted with permission of The American Electroplaters & Surface Finishers Soc. From Reference 27.

These geometric considerations show that the true average thickness is the apparent thickness reduced by a factor R. This is strictly true for vary thin deposits, but becomes less important for thicker plates (27). The direction of the change as a function of thickness depends on the microthrowing power of the solution, and is therefore, specific to the type of solution used (28). Some roughness factors for metals abraded in various ways are presented in Table 2.

Surface roughness also influences mean thickness and spread. An example is shown in Figure 9 for 2.9µm (115/µin) of gold deposited on coarse 0.75µm (30µin) CLA, and polished 0.04µm (1.5µin) CLA, OFHC copper. The data represented by the open circles show that on the polished substrate, 1% of the surface had plate thinner than 2.6µm (101µin), while 99% of the plate was thinner than 3.2µm (127µin). The curve for the rough substrate has a lower mean value and also a larger spread. Its extreme areas are considerably thinner than the thinnest parts of the deposit on the smooth

Table 2: Some Roughness Factors from the Literature

Metal	Treatment	Roughness Factor
Copper	#320 grit paper	4.2
Copper	#800 grit paper	3.5
Stainless steel	#320 grit paper	2.7
Stainless steel	#800 grit paper	2.0
Aluminum	#320 grit paper	3.1
Aluminum	#800 grit paper	2.2
Gold	Coarse crocus	2.5-2.7
Gold	Fine crocus	1.7
Gold	2/0 machine paper	2.6-3.0
Gold	1/0 machine paper	3.7
Gold	2/0 emery cloth	6.0
Aluminum	Mill rolled	3.1
Aluminum	#600 Alundum	2.1
Aluminum	#120 Aloxite	3.0
Aluminum	#240 Aloxite	3.4
Aluminum	#0 paper	9.3
Aluminum	#2/0 paper	17.6
Aluminum	#3/0 paper	19.9

* From reference 27.

surface. For example, 1% of the deposit on the smooth surface is less than 2.6μm (101μin) thick, while for the rough surface 1% of the deposit lies below 1.2μm (47μin). Both had the same apparent thickness of deposit (26). The roughness factor for the coarse surface derived from the deposit thickness measurements was computed to be 1.4 and this agreed with a value of 1.4 using a bent wire to conform to the surface profile (27).

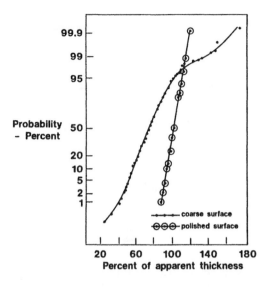

Figure 9: Distribution of thickness measurements made on acid citrate gold plated on coarse and polished OFHC copper. Average thickness—115 microinches. Adapted from Reference 26.

INFLUENCE OF PLATING SOLUTION AND ITS OPERATING PARAMETERS

The porosity in a coating varies with: 1) the concentration of all salts in the solution; 2) the presence of addition agents; 3) the accumulation of aging byproducts; 4) the form of current and the current density used for deposition; 5) the degree of agitation; and 6) the temperature of the solution. Extensive investigations for nickel, copper, gold, cobalt, tin and tin-nickel have verified these general effects which are reviewed in an excellent article by Clarke (21). Figure 10 shows the influence of pulse plating for unbrightened gold on copper. The same three phases that are present for non-pulse plated gold (substrate dominated, transition and coating dominated) are evident but they are displaced downward (22,29).

METHODS TO REDUCE POROSITY

A. Underplates

A convenient and often very effective way to modify the substrate and reduce porosity is to use an underplate. Figure 11 shows that

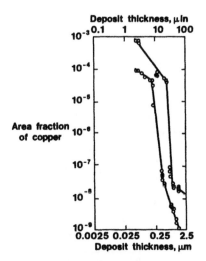

Figure 10: Porosity versus deposit thickness for pulse plated gold on a copper substrate. The curve for an unbrightened gold deposit on a copper substrate (top) is shown for comparison (Also see Figure 5). Adapted from Reference 22.

a sharp reduction in gold plate porosity is achieved with copper underplate of sufficient thickness (26). Another example is Figure 12 which shows the reduction in porosity brought about by using a pure soft gold underplate for a cobalt-hardened gold (30). The curve marked "old" shows the results obtained from a solution which had been replenished several times, while the curve marked "new" is for a fresh solution. The curve marked "duplex" shows results for both solutions with a strike deposit of 13μm. The horizontal dashed line indicates an arbitrary acceptable porosity level for the nitric vapor test used to assess porosity.

B. Crystallographic Orientation

The chapter on structures contains some discussion on crystallographic planes and directions in crystal lattices. Morrissey and his coworkers (22,24,25) effectively utilized this type of information to demonstrate that covering power and rate of pore closure of bright gold deposits on copper are related to the crystallographic orientation of the deposits. The following is extracted from their work.

Gold crystallizes in a face centered cubic structure wherein the most densely packed planes are (111) followed by (200) and then (220) planes. If atoms were added at a constant rate to a crystallite with these

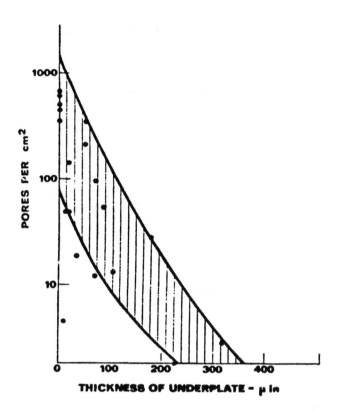

Figure 11: Porosity of 30 microinches of acid citrate gold plated over various thicknesses of acid sulfate copper underplate on OFHC copper substrate. From reference 26. Reprinted with permission of The American Electroplaters & Surface Finishers Soc.

exposed faces, the (111) face would grow at the slowest rate and the (220) face at the fastest rate. Therefore, an electrodeposit with a strong (111) orientation with respect to its surface has its slowest growing crystal planes in the plane of the substrate. This will tend to cause it to grow outward at a much faster rate than it does laterally. By contrast, a deposit with a (220) orientation has fast growing crystal planes in the plane of the substrate so it will tend to grow laterally at a faster rate than outward from the surface (24). Some porosity-thickness results for a series of highly oriented gold deposits on copper, shown in Figure 13, verify these statements. One observation is that the covering power of the various gold deposits decreases with decreasing closeness of atomic packing (111) > (200) > (220). The reason for this is that nucleation on a (111) face lays down more gold atoms than on a (200) or (220) face of equal area. Another observation is that the

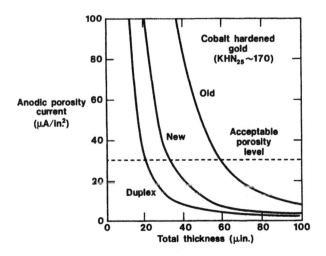

Figure 12: Reduction in porosity achieved by using a pure soft gold underplate for a cobalt-hardened gold. The curve marked "duplex" shows results for both solutions with a strike deposit of 13μm. Adapted from Reference 30.

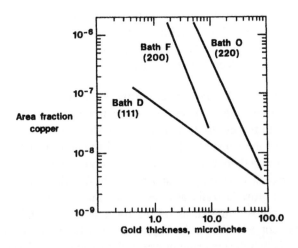

Figure 13: Porosity versus thickness for gold deposits of various preferred orientations. Adapted from Reference 24.

rate of pore closure, which is obtained from the slope of the curve, is greatest for deposits of highest index orientation. Once developed, these high index faces grow laterally at a faster rate than low index faces (24).

Figure 14 shows porosity-thickness results for a highly oriented gold deposit. At 33°C, this deposit shows a high (111) orientation while at 49°C the deposit is (111) preferred, but with appreciable (200) and (220) contributions. The covering power is excellent but the rate of pore closure is slow at 33°C. By contrast, at 49°C the covering power is poorer but the rate of pore closure is quite rapid. By splitting the deposition operation between the two solutions, e.g., nucleating the deposit at 33°C to obtain the best initial coverage, then doing the final plating at 49°C to take advantage of more rapid pore closure, optimum results are obtained in terms of coverage and reduced porosity. This approach has been used in production (24).

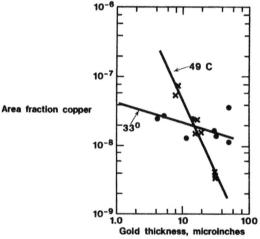

Figure 14: Porosity versus thickness for deposits produced in the same solution at different temperatures. Adapted from reference 24.

C. Deposition Technique—Comparison of Electroplated and Physically Vapor Deposited Films

The porosity of gold thin films depends on the method of deposition. Evaporated and sputtered films are noticeably more porous than plated films. For example, electrodeposits of copper, nickel, and gold developed continuous films at an average thickness of less than 50 Å on copper and nickel substrates, whereas, evaporated films 500 Å or less in thickness were not continuous and showed many holes or channels (31). Figures 15 and 16 show that in two different porosity tests, electrodeposited gold films performed much better than films prepared by physical vapor deposition (32,33).

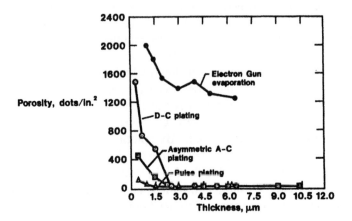

Figure 15: Relationship between porosity and thickness of gold films deposited by electron gun evaporation, asymmetric a-c plating, d-c plating and pulse plating. Substrate material was copper and a nitric acid test was used to measure porosity. Adapted from reference 32.

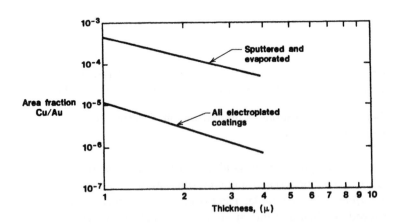

Figure 16: Comparison of porosity of evaporated, sputtered and electrodeposited gold films. Substrate was copper and porosity was measured by an electrochemical test. Adapted from Reference 33.

D. Hot Isostatic Pressing (HIP)

Porosity can be suppressed by hot isostatic pressing (HIP), a heat treatment under high pressure. However, because of the high temperature used, e.g., 550°C under 21 kg/mm² (30 kpsi) nitrogen pressure for 2 hours for copper, the process results in grain growth (34). HIP has also been used to suppress Kirkendall porosity formation (35).

POROSITY TESTING

The purpose of this section is to describe porosity tests that have been used. For an extremely comprehensive review on this subject see reference 21; references 1, 36 and 37 also provide good detail. Table 3 lists the various types of porosity tests.

Table 3: Porosity Tests

Chemical
    color change
    chemical analysis

Gas
    sulfur dioxide
    nitric acid vapor

Electrolytic
    anodic current plot
    electrographic printing

Microscopic
    optical metallography
    scanning electron microscopy
    transmission electron microscopy
        a - diffraction contrast
        b - phase contrast

Density measurements

Detached Coatings
    gas flow
    visual and photographic

Porosity tests may be broken down into four categories:

1.   Pore detection tests which make pore sites visible for inspection and counting. This includes detection in situ by producing visible corrosion product, detection by radiography, and examination of coatings detached from their substrate.

2.   Porosity index tests which provide a direct numerical measure.

3.   Microscopic techniques which permit direct observation of both isolated volds and through pores.

4.   Density measurements which provide an indirect measure of both isolated voids and through pores.

A. Chemical Tests

Color tests involve use of a reagent which causes a color change in the presence of corrosion products from the substrate, forming a distinctive spot at each pore. An example of this is the classic ferroxyl test for steel which uses a solution of sodium chloride and potassium ferricyanide (38). Chemical analysis tests rely on quantitative determination of substrate corrosion products by chemical analysis, such as the use of an ammonia- ammonium persulfate solution which attacks a copper substrate but not a gold overcoat, so the amount of dissolved copper is a measure of the porosity of the coating (39).

B. Gas Exposure Tests

Gas exposure porosity tests offer two potential advantages over liquid immersions tests: 1) the gas has a better ability to penetrate small pores, since surface tension effects could inhibit such penetration by bulk liquids; and 2) many gaseous porosity tests simulate pore corrosion mechanisms that may actually occur in service (40). Two gas tests that have been used include the humid, 10% sulfur dioxide test and the nitric acid vapor test (41). The 10% sulfur dioxide test atmosphere is generated by mixing a 50% solution of concentrated sulfuric acid with a 20% (by weight) solution of sodium thiosulfate. The ratio of the sulfuric acid solution to that of the thiosulfate is usually 1:4. The nitric acid vapor test relies on corrosive vapor produced directly from concentrated nitric acid that has been placed in the bottom of the test vessel. This test is limited to gold and platinum coatings (40).

C. Electrolytic Techniques

Electrolytic techniques offer the opportunity for rapid and relatively nondestructive means of determining porosity with high sensitivity (42,43). Ogburn (44) lists three types of electrolytic tests wherein the specimen is immersed in an electrolyte with an auxiliary electrode and a reference electrode: 1) the current is measured while the specimen is made anodic; 2) the anodic polarization curve slope is determined; or 3) a measurement is made of the corrosion potential.

Good examples of the quantitative type of data that can be obtained with electrochemical porosity measurements are shown in Figures 5, 10, 13, 14, and 16, which present corrosion potential measurements made with gold plated copper samples in 0.1 M ammonium chloride electrolyte. The corrosion potential of the sample is related to the exposed basis metal area fraction and this relationship affords a convenient and sensitive means for determining porosity.   In addition to use of this test to provide quantitative data for gold on copper (33,42), it has been used for nickel on uranium (45,46) gold on tungsten (47), and tin on steel (48).

Electrographic porosity testing, also referred to as electrography or electrographic printing, involves use of chemically impregnated dye-transfer paper which is pressed firmly against the surface to be examined. Current is passed from the specimen which is anodic to an inert cathode at a fixed current density for a specified time. Cations from the substrate are formed at pores or cracks in the protective coating under the influence of the applied potential.   These cations enter the gelatinized surface of the dye-transfer paper and react with appropriate chemicals to form soluble complexes or colored precipitates. Pores appear as colored dots while cracks appear as colored lines on the print (49). This test is quick, reproducible, suitable for on-line testing and provides a print which can saved for future reference (50). A schematic of the test set-up is shown in Figure 17.   Table 4 lists chemicals used for various coatings and substrates (51). Some examples of use of electrography to measure porosity of electrodeposited coatings include gold on copper and nickel substrates (50,52) and chromium on various substrates (49). Figures 18a and 18b are electrographic prints of 1.5 μm thick decorative chromium on a nickel and copper plated zinc base die casting illustrating the effect of heat cycling three times to 80°C. Electrography has also been used to detect pinholes in thin dielectric films (53), polymer films on metallic substrates (54,55) and chemically vapor deposited coatings on cemented carbide substrates (56).

Porosity testing in a gelled medium is a special form of electrographic testing (electrography) which eliminates many of the disadvantages associated with pressure electrography (50,57). In this test,

270   Electrodeposition

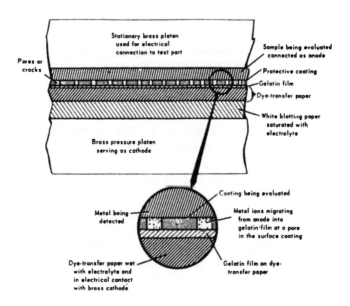

Figure 17: Detail of electrographic printing method illustrating principle of operation. From reference 49. Reprinted with permission of The American Electroplaters & Surface Finishers Soc.

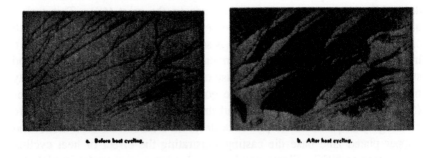

a. Before heat cycling.   b. After heat cycling.

Figure 18: Electrographic prints of 0.06 mil thick chromium on a nickel and copper plated zinc-base die casting illustrating the effect of heat cycling three times to 80°C. From Reference 49. Reprinted with permission of The American Electroplaters & Surface Finishers Soc.

Table 4: Reagents Used In Electrographic Tests

Deposit	Solution
Gold on copper, silver on copper	Potassium ferricyanide(brown spots)
Tin on iron	Potassium ferricyanide(blue spots)
Tin on brass	Antimony sulfide plus phosphoric acid (brown spots)
Gold on nickel	a-Ammoniacal dimethylglyoxime and sodium chloride (red spots), b-Alcoholic dimethylglyoxime and sodium chloride (red spots)
Chromium on nickel	Dimethylglyoxime (pink spots)
Copper on iron	Dimethylglyoxime (deep cherry red spots)
Nickel on steel	Sodium chloride plus hydrogen peroxide (rust spots)
Gold on bronze	Ammonium phosphomolybdate
Nickel on copper or brass	Sodium nitrate solution, developed later with potassium ferricyanide solution plus acetic acid
Tin on lead	Ammonium acetate and potassium chromate
Gold, platinum or iridium on molybdenum	Potassium bisulfate developed later with potassium ferricyanide (dark brown spots)
Zinc or cadmium on steel	Half-normal sodium hydrosulfide (black spots)
Zinc on iron	Potassium ferrocyanide and magnesium sulfate
Deposits on iron or copper anodes	Potassium ferricyanide solution
Deposits on silver anodes	Potassium bichromate solution
Thin polymer films or dielectrics	Benzidine (references 54 and 55)

*Unless otherwise noted all data in this table are from reference 51. Additional details on solution formulations can be found in reference 49.

the specimen is made the anode in a cell containing a solid or semisolid electrolyte of gelatin, conducting salts, and an indicator. Application of current results in the migration of base metal ions through continuous pores. The cations react with the indicator giving rise to colored reaction products at pore sites, and these may be counted through the clear gel. Table 5 lists various electrolyte solutions and their resulting indicator colors. This method is suitable for coatings commonly used on electrical contacts, e.g., gold on substrates of silver, nickel, copper and its alloys, and for coatings of 95% or more of palladium on nickel, copper and its alloys.

Gel-bulk electrography offers a number of advantages over paper electrography: pressure control difficulties are eliminated, special mounting or clamping devices are not required, exposed surfaces of any geometry, even complex shapes can be tested, and the exact location of pores can be observed without having to index a piece of test paper which shows a mirror image of the test part. Also, the gel method appears to be more sensitive in that smaller pores can be detected and current flow is much smaller, reducing the possibility of producing pores by the test procedure (57).

Table 5: Gel Porosity Testing Solutions

Test for	Electrolyte (Aqueous)	Indicator	Indicating Color	Comments
Copper[a]	4% sodium carbonate + 1% sodium nitrate	saturated solution of rubeanic acid in ethanol	dark olive green	also detects nickel, cobalt
Copper	4% sodium carbonate + 1% sodium nitrate	7.5% potassiuim ferrocyanide in water	brown	-----
Nickel[a]	4% sodium carbonate + 1% sodium nitrate	saturated solution of rubeanic acid in ethanol	blue-blue violet	also detects copper, cobalt
Nickel	5% ammonium hydroxide	saturated solution of dimethylglyoxime in ethanol	pink	----------
Silver[ab]	0.2 molar nitric acid	1% glacial acetic + 5% sodium dichromate	red	solution must be free of halogens
Silver[b]	0.2 molar nitric acid	saturated solution of rhodanine in ethanol	red-red violet	solution must be free of halogens

[a]Preferred Test
[b]Not suitable for palladium overplates

* From reference 57

D. Microscopic Techniques

Microscopic techniques, which include optical metallography (OM), scanning electron microscopy (SEM) and transmission electron microscopy (TEM), permit direct observation of both isolated voids and through pores (1). Voids intersecting the film surface can be detected by OM and SEM while TEM is more useful in examining both small and large voids which can be isolated inside a film or located on the film surface. With use of the defocus contrast technique (phase contrast) in TEM (58), high density (about 1 x 10^{17}/cm^3) very small (approximately 10Å) voids have been found in thin films prepared by evaporation, sputtering and electrodeposition (3). Of interest is that this demonstrates that voids are generally formed in thin films irrespective of the film preparation method as long as the deposition process involves a phase transition from the vapor to the solid state (1).

REFERENCES

1. S. Nakahara, "Porosity in Thin Films", *Thin Solid Films*, 64, 149 (1979)

2. K.C. Joshi and R.C. Sanwald, "Annealing Behavior of Electrodeposited Gold Containing Entrapments", *J. Electron. Materials*, 2, (4), 533 (Nov 1973)

3. S. Nakahara, "Microporosity Induced by Nucleation and Growth Processes in Crystalline and Non-Crystalline Films", *Thin Solid Films*, 45, 421 (1977)

4. J.M. Leeds, "A Survey of the Porosity in Gold and Other Precious Metal Electrodeposits", *Trans. Inst. of Metal Finishing*, 47, 222 (1969)

5. R.G. Baker, C.A. Holden and A. Mendizza, "Porosity in Electroplated Coatings-A Review of the Art With Respect to Porosity Testing", *50th Annual Technical Proceedings*, American Electroplaters Society, 61 (1963)

6. S.M. Garte, "Porosity", *Gold Plating Technology*, F.H. Reid and W. Goldie, Editors, Electrochemical Publications Ltd., 295 (1974)

7. D. Altura, "Postplating Embrittlment-Behavior of Several Cadmium Deposits", *Metal Finishing*, 72, 45 (Sept 1974)

8. J.G. Rinker and R.F. Hochman, "Hydrogen Embrittlement of 4340 Steel as a Result of Corrosion of Porous Electroplated Cadmium", *Corrosion*, 28, 231 (1972)

9. R.C. Sanwald, "The Characteristics of Small Gas Filled Voids in Electrodeposited Gold", *Metallography*, 4, 503 (1971)

10. R.J. Bourcier, D.A. Koss, R.E. Smelser and O. Richmond, "The Influence of Porosity on the Deformation and Fracture of Alloys", *Acta Metall.*, 34, 2443 (1986)

11. M. Myers and E.A. Blythe, "Effects of Oxygen, Sulphur, and Porosity on Mechanical Properties of Cast High-Purity Copper at 950 C", *Metals Technology*, 8, 165 (May 1981)

12. R.R. Zito, "Failure of Reflective Metal Coatings by Cracking", *Thin Solid Films*, 87, 87 (1982)

13.   S. Nakahara and Y. Okinaka, "On the Effect of Hydrogen on Properties of Copper", *Scripta Metallurgia*, 19, 517 (1985)

14.   A. Staudinger and S. Nakahara, "The Structure of the Crack Network in Amorphous Films", *Thin Solid Films*, 45, 125 (1977)

15.   R.L. Zeller, III and U. Landau, "The Effect of Hydrogen on the Ductility of Electrodeposited Ni-P Amorphous Alloys", *J. Electrochem. Soc.*, 137, 1107 (1990)

16.   T.W. Tomaszewski, L.C. Tomaszewski and H. Brown, "Codeposition of Finely Dispersed Particles With Metals, *Plating* 56, 1234 (1969)

17.   C.L. Faust and W.H. Safranek, "The Electrodeposition of Porous Metal", *Trans. Inst. of Metal Finishing*, 31, 517 (1954)

18.   P.R. Coronado, "Porous Nickel Deposits by Codeposition of Plastic Powders, *SCL-DR-67-64*, Jan 1968, Sandia Laboratories, Livermore, CA

19.   Y. Tazaki et al., British Patent Application, GB 2,206,896A, Jan 18, 1989

20.   A. Kutzelnigg, "The Porosity of Electrodeposits, Causes, Classification and Assessment", *Plating* 48, 382 (1961)

21.   M. Clarke, "Porosity and Porosity Tests", *Properties of Electrodeposits: Their Measurement and Significance*, R. Sard., H. Leidheiser, Jr., and F. Ogburn, Editors, The Electrochemical Soc., 122 (1975)

22.   R.J. Morrissey, "Porosity and Galvanic Corrosion in Precious Metal Electrodeposits", *Electrochemical Techniques for Corrosion Engineering*, R. Baboian, Editor, National Association of Corrosion Engineers, Houston, Texas (1985)

23.   D.R. Gabe, "Metallic Coatings for Protection", Chapter 4 in *Coatings and Surface Treatment for Corrosion and Wear Resistance*, K.N. Strafford, P.K. Datta and C.G. Googan, Editors, Ellis Horwood Limited (1984)

24.   A.M. Weisberg, H. Shoushanian and R.J. Morrissey, "Methods of Reducing Porosity in Gold Deposits", *Proceedings Design and Finishing of Printed Wiring and Hybrid Circuits Symposium*, American Electroplaters Soc. (1976)

25.   R.J. Morrissey and A.M. Weisberg, "Some Further Studies on Porosity in Gold Electrodeposits", *Corrosion Control By Coatings*, H. Leidheiser, Jr., Editor, Science Press (1979)

26.   S.M. Garte, "Porosity of Gold Electrodeposits: Effect of Substrate Surface Structure", *Plating* 55, 946 (1968)

27.   S.M. Garte, "Effect of Substrate Roughness on the Porosity of Gold Electrodeposits", *Plating* 53, 1335 (1966)

28.   O. Kardos and D. G. Foulke, *Advances in Electrochemistry and Electrochemical Engineering*, P. Delahay and C.W. Tobias, Editors, Vol 2, (1962)

29.   J. Mazia and D.S. Lashmore, "Electroplated Coatings", *Metals Handbook Ninth Edition, Volume 13, Corrosion*, ASM International, Metals Park, Ohio (1987)

30.   R.G. Baker, H.J. Litsch and T.A. Palumbo, "Gold Electroplating, Part 2-Electronic Applications", Illustrated Slide Lecture, American Electroplaters Soc.

31.   K.R. Lawless, "Growth and Structure of Electrodeposited Thin Metal Films", *J. Vac. Sci. Technol.*, 2, 24 (1965)

32.   D.L. Rehrig, "Effect of Deposition Method on Porosity in Gold Thin Films", *Plating* 61, 43 (1974)

33.   J.W. Dini and H.R. Johnson, "Optimization of Gold Plating for Hybrid Microcircuits, *Plating & Surface Finishing*, 67, 53 (Jan 1980)

34.   J.C. Farmer, H.R. Johnson, H.A. Johnsen, J.W. Dini, D. Hopkins and C.P. Steffani, "Electroforming Process Development For the Two-Beam Accelerator", *Plating & Surface Finishing*, 75, 48 (March 1988)

35.   I.D. Choi, D.K. Matlock and D.L. Olson, "Creep Behavior of Nickel-Copper Laminate Composites With Controlled Composition Gradients", *Metallurgical Transactions* A 21A, 2513 (1990)

36.   "Selection of Porosity Tests for Electrodeposits and Related Metallic Coatings", *ASTM B765-86* (1986), American Society for Testing and Materials

37.   F.J. Nobel, B.D. Ostrow and D.W. Thompson, "Porosity Testing of Gold Deposits", *Plating 52*, 1001 (1965)

38.   W.H. Walker, *J. Ind. Eng. Chem.*, 1, 295 (1909)

39.   M.S. Frant, "Porosity Measurements on Gold Plated Copper", *J. Electrochem. Soc.*, 108, 774 (1961)

40.   S.J. Krumbein and C.A. Holden, Jr., "Porosity Testing of Metallic Coatings", *Testing of Metallic and Inorganic Coatings, ASTM STP 947*, W.B. Harding and G.A. DiBari, Eds., American Society for Testing and Materials, 193 (1987)

41.   "Porosity in Gold Coatings on Metal Substrates by Gas Exposure", ASTM B735-84 (1984), American Society for Testing and Materials

42.   R.J. Morrissey, "Electrolytic Determination of Porosity in Gold Electroplates, I. Corrosion Potential Measurements", *J. Electrochem. Soc.*, 117, 742 (1970)

43.   R.J. Morrissey, "Electrolytic Determination of Porosity in Gold Electroplates, II. Controlled Potential Techniques", *J. Electrochem. Soc.*, 119, 446 (1972)

44.   F. Ogburn, "Methods of Testing", *Modern Electroplating, Third Edition*, F.A. Lowenheim, Editor, Wiley-Interscience, (1974)

45.   L.J. Weirick, "Electrochemical Determination of Porosity in Nickel Electroplates on a Uranium Alloy", *J. Electrochem. Soc.*, 122, 937 (1975)

46.   W.C. Dietrich, "Potentiometric Determination of Percent Porosity in Nickel Electroplates on Uranium Metal", *Proceedings of Second AES Plating on Difficult-to-Plate Materials Symposium*, American Electroplaters Society (March 1982)

47.   F.E. Luborsky, M.W. Brieter and B.J. Drummond, "Electrolytic Determination of Exposed Tungsten on Gold Plated Tungsten", *Electrochimica Acta*, 17, 1001 (1972)

48.   I. Notter and D. R. Gabe, "The Electrochemical Thiocyanate Porosity Test for Tinplate", *Trans. Inst. Metal Finishing*, 68, 59 (May 1990)

49.   H.R. Miller and E.B. Friedl, "Developments in Electrographic Printing", *Plating 47*, 520 (1960)

50.   H.J. Noonan, "Electrographic Determination of Porosity in Gold Electrodeposits", *Plating 53*, 461 (1966)

51.   F. Altmayer, "Simple QC Tests for Finishers", *Products Finishing*, 50, 84 (Sept 1986)

52.   "Porosity in Gold Coatings on Metal Substrates by Paper Electrography", *ASTM B741-85* (1985), American Society for Testing and Materials

53.   J.P. McCloskey, "Electrographic Method for Locating Pinholes in Thin Silicon Dioxide Films", *J. Electrochem. Soc.*, 114, 643 (1967)

54.   S.M. Lee, J.P. McCloskey and J.J. Licari, "New Technique Detects Pinholes in Thin Polymer Films", *Insulation*, 40 (Feb 1969)

55.   S.M. Lee and P.H. Eisenberg, "Improved Method for Detecting Pinholes in Thin Polymer Films", *Insulation*, 97 (August 1969)

56.   A. Tvarusko and H. E. Hintermann, "Imaging Cracks and Pores in Chemically Vapor Deposited Coatings by Electrographic Printing", *Surface Technology*, 9, 209 (1979)

57.   F.V. Bedetti and R.V. Chiarenzelli, "Porosity Testing of Electroplated Gold in Gelled Media", *Plating 53*, 305 (1966)

58.   S. Nakahara and Y. Okinaka, "Transmission Electron Microscopic Studies of Impurities and Gas Bubbles Incorporated in Plated Metal Films", Chapter 3 in *Properties of Electrodeposits: Their Measurement and Significance*, R. Sard, H. Leidheiser, Jr., and F. Ogburn, Editors, The Electrochemical Soc., Pennington, NJ, (1975)

9
STRESS

INTRODUCTION

Stresses which remain in components following production may be so high that the total of operating and residual stress exceeds the material's strength. Among the most famous examples of failures due to residual stresses were the all-welded Liberty ships of World War II (1). Over 230 of these ships were condemned because of fractures arising from failures below the design strength; these were directly due to the existence of unrelieved residual stresses set up during the welding. One T-2 tanker, the "Schenectady", had the unenviable distinction of breaking in half while being fitted out at the pier in calm seas during mild weather and without ever having "gone to sea". Investigation showed that the maximum bending moments from the loading at the time it broke up were under one-half those allowed for in the design; it had failed because it was severely over-stressed by residual stresses alone (1). Although examples from coatings are not as dramatic, residual stresses introduced as a result of the deposition process can create problems.

A residual stress may be defined as a stress within a material which is not subjected to load or temperature gradients yet remains in internal equilibrium. Residual stresses in coatings can cause adverse effects on properties. They may be responsible for peeling, tearing, and blistering of the deposits; they may result in warping or cracking of deposits; they may reduce adhesion, particularly when parts are formed after plating and may alter properties of plated sheet. Stressed deposits can be considerably more reactive than the same deposit in an unstressed state. This point is clearly shown in Figure 1 which compares the reaction of highly stressed and

Figure 1:   Reaction of rhodium deposits with different stresses upon exposure to nitric acid solution. Deposit on the left had a tensile stress of 690 MPa while that on the right had a compressive stress of 17 MPa. From reference 2. Reprinted with permission of ASM International.

slightly compressively stressed rhodium deposits to nitric acid. Silver coupons were plated with 5 μm (0.2 mil) thick rhodium deposits. In one case the stress in the rhodium was 690 MPa (100,000 psi) tensile while the other was 17 MPa (2500 psi) compressive. Once released from the restraining substrate by dissolution of the silver in nitric acid, the highly stressed rhodium exhibited catastrophic failure (2).

Occasionally stress may serve a useful purpose. For example, in the production of magnetic films for use in high speed computers, stress in electrodeposited iron, nickel, and cobalt electrodeposits will bring about preferred directions of easy magnetization and other related effects (2).

THERMAL, RESIDUAL, AND STRESS DURING SERVICE

Two kinds of stress exist in coatings: differential thermal stress and, residual or intrinsic stress (3). Differential thermal stresses can be calculated. For example, assuming a twofold difference in coefficient of expansion between the basis metal and the coating (the differences are usually smaller), a temperature change of 100°C will produce stresses on the order of 69 MPa (10,000 psi) to 207 MPa (30,000 psi) (4). An electroless nickel deposit will shrink about 0.1 percent when cooled from a plating solution temperature of 90°C to ambient temperature (5)(6). Depending on the thermal coefficient of expansion of the substrate, the stress induced in the coating can be either tensile or compressive. Heat treating electroless nickel deposits above 250°C increases the tensile stress due to the volume shrinkage that occurs during nickel phosphide precipitation and nickel crystallization (5). More information on this is included in the chapter on Structure.

Besides differential thermal stress and stress from the coating process, an added stress can be introduced during use of the plated part. An example is gold plated spectacle frames. One source of corrosive attack on plated surfaces is the formation of cracks, thereby exposing the substrate. Corrosion of spectacle frames can occur due to attack of perspiration through cracks which develop if the sum of the tensile stresses in the metal exceeds the tensile strength of the plating. In addition to thermal and residual stresses, a stress component can result from service usage. Bending or twisting of the plated spectacle frames can cause tensile stresses in the convex layers (7). Table 1 summarizes the stresses that might occur with these parts. When the combined tensile stresses exceed the tensile strength of the plating, cracks can develop and these expose the basis metal to corrosive attack. The data in Table 1 indicate that in this situation there seems to be no danger of cracking even if all three effects take place simultaneously since the tensile strength of the gold is not exceeded (7).

Table 1. Stress Data for a Gold-Nickel Deposit on a Spectacle Framed)[a]

Inner stress of plating	46 MPa	(6670 psi)
Temperature gradient of 10°C	26 MPa	(3770 psi)
Bending radius of 5 cm	100 MPa	(14500 psi)
Sum	172 MPa	(24940 psi)
Tensile strength of gold	200 MPa	(29000 psi)

[a] = From Reference 7

Table 2 provides data on the relative magnitude of stresses in electrodeposits. It's interesting to note that there is an apparent relationship between stress and melting point with the transition metals exhibiting the highest tensile stresses (2). Tensile stress (+) causes a plated strip to bend in the direction of the anode; this type of bending is met when the deposit is distended and tends to reduce its volume. A plated strip that bends away from the anode is compressively stressed (-); this type of bending occurs when the deposit is contracted and tends to increase in volume (8). The data in Table 2 can be noticeably influenced by additives and this is discussed in the chapter on Additives.

Table 2: Stress Data for Some Electrodeposited Metals (1)

Deposit	Melting Point (°C)	Stress(2) MPa	psi
Cadmium	321	-3.4 to -20.7	-500 to -3000
Zinc	420	-6.9 to -13.8	-1000 to -2000
Silver	961	±13.8	±2000
Gold	1063	-3.4 to 10.3	-500 to 1500
Copper	1083	13.8	2,000
Nickel	1453	68.9	10,000
Cobalt	1495	138	20,000
Iron	1537	276	40,000
Palladium	1552	413	60,000
Chromium	1875	413	60,000
Rhodium	1966	689	100,000

1. From reference 2.
2. Minus values represent compressive stress.

INFLUENCE OF RESIDUAL STRESS ON FATIGUE

Electrodeposits have been known to reduce the fatigue strength of plated parts. The reasons for this include: 1) hydrogen pickup resulting from the cleaning/plating process, 2) surface tensile stresses in the deposits, and 3) lower strength of the deposits compared to the basis metal leading to cracks in the deposit which subsequently propagate through to the base metal. A wealth of information on the influence of electrodeposits on fatigue strength can be found in reference 9. The discussion in this chapter will focus only on the influence of stress in electrodeposits on fatigue strength.

A general rule of thumb is that tensile stresses in the deposits are deleterious, and the higher the stress the worse the situation in regards to fatigue strength of the substrate. It is also important to realize that the strength of the steel also affects the amount of reduction in fatigue strength obtained after electrodeposition. Data in Table 3 present information for a variety of deposits on two steels, SAE 8740 and SAE 4140. In all cases, a reduction in endurance limit was obtained as a function of increasing residual stress in the deposit.

Table 3: Influence of Residual Stress in Various Electrodeposited Coatings on Fatigue Properties of SAE 8740 and SAE 4140 Steels

SAE 8740 (AMS 6322) Steel (1)

Electrodeposit	Deposit Residual Stress		Endurance Limit (10^7cycles)	
	MPa	psi	MPa	psi
Sulfamate nickel	-41	-6,000	621	90,000
	21	3,000	614	89,000
	83	12,000	552	80,000
	131	19,000	483	70,000
Sulfamate nickel-	55	8,000	628	91,000
cadmium	76	11,000	531	77,000
	110	16,000	483	70,000
Watts nickel-	173	25,000	476	69,000
cadmium	214	31,000	386	56,000

SAE 4140 Steel (2)

None	---	---	752	109,000
Lead	0	0	725	105,000
Bright nickel	-21	-3,000	587	85,000
Watts nickel	173	25,000	310	45,000

1.   Data for SAE 8740 are from reference 10. The steel was hardened and tempered to Rockwell C 37-40 and had a tensile strength of 1240 MPa (180,000 psi). All plated with coatings were 7.5 to 12.5 µm (0.3 to 0.5 mil) thick.

2.   Data for 4140 are from reference 11. The tensile strength of the steel was 1456 MPa (211,000 psi). Thickness of all deposits was 25 µm (1 mil).

Typically, chromium deposits highly stressed in tension reduce the fatigue strength of steel substrates to a greater degree than deposits with less stress (12-14). Compressively stressed chromium deposits reduce the fatigue strength of steel substrates very slightly or not at all, depending on the strength of the steel and degree of compressive stress.

Shot peening before plating to induce compressive stress in the surface layers of the steel can help reduce the fatigue loss from subsequent plating. Steel with a tensile strength of 1380 Mpa (200,000 psi) which had been reduced 47 percent in fatigue strength by chromium plating, was reduced only 10 percent in fatigue strength when it was shot peened before plating. In another case, a steel with a tensile strength of 1100 MPa (160,000 psi) reduced 40 percent in fatigue strength by chromium plating was reduced only about 5 percent in fatigue strength when it was shot peened before plating (12). The federal chromium plating specification, QQ-C-320 calls for parts that are designed for unlimited life under dynamic loads to be shot peened and baked at 190°C (375°F) for not less than three hours.

HOW TO MINIMIZE STRESS IN DEPOSITS

There are a variety of steps that can be taken to minimize stress in deposits:

-choice of substrate
-choice of plating solution
-use of additives
-use of higher plating temperatures

Influence of Substrate

Typically, with most deposits, there is a high initial stress associated with lattice misfit and with grain size of the underlying substrate. This is followed by a drop to a steady state value as the deposit increases in thickness. With most deposits this steady state value occurs in the thickness regime of 12.5 - 25 μm (0.5 - 1.0 mil).   Atomic mismatch between the coating and substrate is a controlling factor with thin deposits. For example, when gold is plated on silver, the influence of mismatch is almost absent because the difference in the interatomic spacings of gold and silver is only 0.17%. This is quite different for copper and silver since the difference in this case is about, 13% (15).

A curve showing the relationship of stress in nickel deposited on different copper substrates is shown in Figure 2. The initial high stress is due to lattice misfit and grain size of the underlying metal. With fine grained substrates, the maximum stress is higher and occurs very close to the interface. As the thickness increases the stress decreases to a steady state value, the finer the substrate grain size, the more rapid this descent(2). The influence of the substrate on stress is also shown in Figure 3 which is a plot of stress in electroless nickel coatings on a variety of substrates (aluminum, titanium, steel, brass and titanium) as a function of phosphorus content. Besides showing that the substrate has a very distinct influence on stress due to lattice and coefficient of thermal expansion mismatches, Figure 3 also shows that for each substrate a deposit with zero stress can be obtained by controlling the amount of phosphorus in the deposit (16).

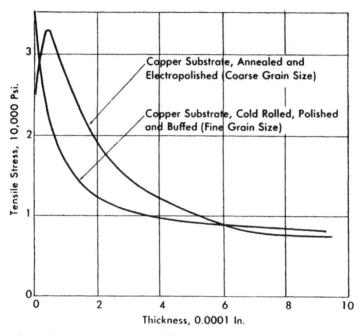

Figure 2: Effect of grain size and deposit thickness on tensile stress in nickel deposited from a sulfamate solution at room temperature. From reference 2. Reprinted with permission of ASM International.

In terms of adhesion, the ideal case which would provide a true atomic bond between the deposit and the substrate is that wherein there is epitaxy or isomorphism (continuation of structure) at the interface. Although this often occurs in the initial stages of deposition, it can only remain throughout the coating when the atomic parameters of the deposit and the substrate are approximately the same. Since the stress which develops at the beginning of the deposition process, is in actuality a measure of bond strength, poor bonding shows up significantly in stress determinations. This is shown in Figure 4 for a nickel deposit on poorly cleaned and properly cleaned 304 stainless steel. The poorly cleaned substrate had been allowed

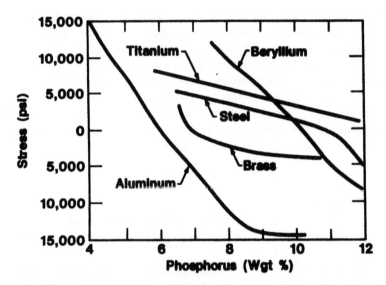

Figure 3: Stress in electroless nickel as a function of phosphorus content for metals with a high expansion coefficient (aluminum and brass) and a low expansion coefficient (steel, beryllium and titanium). Adapted from reference 5.

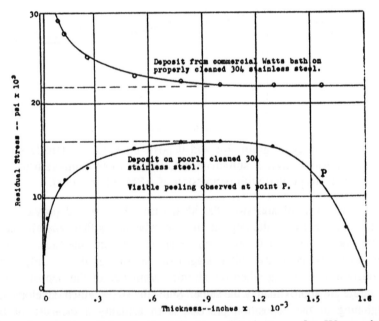

Figure 4: Effect of bond strength on residual stress for Watts nickel deposits on 304 stainless steel. From reference 17. Reprinted with permission of Metal Finishing.

to dry in air prior to nickel plating so that any oxide film destroyed by the cleaning process could reform, while the properly cleaned substrate was not dried prior to immersion in the nickel plating solution. The deposit on the poorly cleaned substrate separated with a light pull while that on the properly cleaned substrate could not be removed even with the use of a knife blade. Figure 4 shows that the poor bonding was reflected by low initial stress values, unlike the typically high values seen when good adhesion was present (17).

Influence of Plating Solution

The type of anion in the plating solution can leave a marked influence on residual stress as shown in Table 4 for nickel deposits produced in different solutions. Sulfamate ion provides nickel deposits with the lowest stress, followed by bromide which also reduces pitting (2). Not all plating solutions offer this wide range of anions capable of providing acceptable deposits but this option should not be ignored when looking for a deposit with low stress.

Table 4: Influence of Anion on Residual Stress

In Nickel Deposits (a)

Solution Anion	Residual Stress	
	MPa	psi
Sulfamate	59	8,600
Bromide	78	11,300
Fluoborate	119	17,200
Sulfate	159	23,100
Chloride	228	33,000

a.   These data are from reference 2. Deposit thickness was 25 μm (1 mil), temperature was 25°C, and current density 323 amp/sq dm. All solutions contained 1 M nickel, 0.5 M boric acid, pH was 4.0 and the substrate was copper.

Influence of Additives

There are numerous additives, particularly organic, which have a marked effect on the stress produced in deposits. In fact, additives are so important in influencing properties of deposits that an entire chapter is devoted to this topic. An example of the influence of additives on stress in nickel deposits is presented here for illustration purposes.

Small quantities (0.01 to 0.1 g/l) of most sulfur bearing compounds rapidly reduce stress in nickel deposits. All oxidation states except the plus six oxidation state found in the stable sulfate provide this effect illustrated in Figure 5. Sulfur in less stable compounds in the plus six oxidation state, such as aryl sulfonate and saccharin also lower stress compressively (18). In all cases where a sulfur compound reduces internal stress compressively, the resulting deposit is brighter. In fact, the observation that a nickel sulfamate solution is starting to produce a bright deposit is a good indicator that the anodes are not functioning properly Polarized or inert anodes in a nickel sulfamate solution result in formation of azodisulfonate which is an oxidation product of sulfamate and a major source of stress-reducing sulfur in the deposit (19). The stress reducing azodisulfonate can be removed by hydro-lysis, on warming, or by a conventional peroxide carbon treatment (19). Besides serving as a stress reducer, sulfur in nickel deposits also exerts a strong influence on notch sensitivity and hardness (20) and can also embrittle deposits at high temperature (21). This is discussed in more detail in the

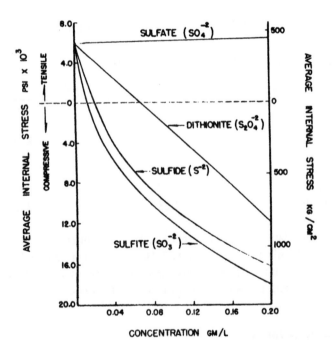

Figure 5: Effects of different forms of sulfur on internal stress in nickel sulfamate deposits. From reference 18. Reprinted with permission of The American Electroplaters & Surface Finishers Soc.

chapter on Properties. Contaminants which find their way into plating solutions by accident or because of careless plating operations can also noticeably influence stress and this should also be kept in mind.

Influence of Plating Solution Temperature

Increasing the plating solution temperature can reduce stress and this is shown in Table 5 for a nickel sulfamate solution (22). Figure 6 visually shows the influence of temperature on stress comparing nickel sulfamate deposits produced at 15°C and 40°C. The deposit produced at 15°C buckled severely due to its high tensile stress while that produced at 40°C showed no deformation.

Table 5: Variation of Residual Stress With Temperature

For a Nickel Sulfamate Solution (a)

Solution Temperature	Deposit Stress	
°C	MPa	psi
1.4	410	59,500
10.5	186	27,000
18.0	91	13,200
25.0	59	8,500
40.0	17	2,500

(a) These data are from reference 22. The plating current density was 323 amp/sq dm.

STRESS MEASUREMENT

There are a variety of techniques used to measure stress in deposits and these are well documented in the literature (8)(23-26). The following discussion on this topic is not meant to be all-inclusive on stress measurement but rather an overview of the more commonly utilized methods and some that offer promise for the future. A listing of stress measurement techniques includes the following:

-rigid or flexible strip
-spiral contractometer
-stresometer
-X-ray
-strain gauge
-dilatometer
-hole drilling
-holographic interferometry

Figure 6: Influence of stress on deposits produced in nickel sulfamate solution at 40°C and 15°C.

Rigid or Flexible Strip

This method is based on plating one side of a long, narrow metal strip (23)(24)(26). The back side of the strip is insulated and one end is clamped while the other is free to deflect either during the plating operation or afterwards (Figure 7). For deposits in tension, the free end deflects towards the anode, while compressive stress is indicated by deflection of the free end of the strip away from the anode.

If one end of the strip is not constrained during plating, the deflection can be recorded by attaching, a pointer or a light source which moves over the scale. This allows for determination of stress as a function

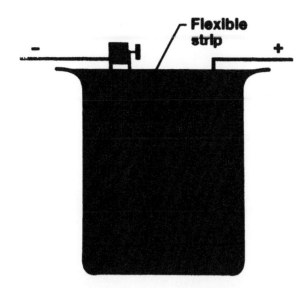

Figure 7: Flexible strip method for measuring residual stress. The cathode can also be restrained during plating. From reference 26.

of thickness. When the strip is constrained during deposition, only the final average stress can be determined (23).

The Stoney formula (27) is used to calculate stress. The variations introduced in this equation over the years to account for the effects of deposit thickness, modulus of elasticity and temperature are covered in detail by Weil (23)(24).

A strip partially cut into eight sections so that each could deflect independently of the others has been used in a Hull cell to obtain data on the effect of current density on stress in one experimental run (28). Another version of the rigid strip principle is shown in Figure 8. During plating, opposite sides of a two legged strip are plated and the resulting deposit causes the strip to spread apart. Deflection is easily measured using the scale shown in Figure 8 (29).

Spiral Contractometer

This instrument, developed by Brenner and Senderoff (30) consists of a strip wound in the shape of a helix and rigidly anchored at one end (Figure 9). The other end is free to move but as it does it actuates a pointer on the dial of the instrument. After calibration with a known force, the stress can be determined from the angle of rotation of the pointer. Compressive residual stress causes the helical strip to unwind while a tensile stress winds it tighter. Over the years a number of modifications have been made to the spiral contractometer technique and these are discussed in detail by Weil (24).

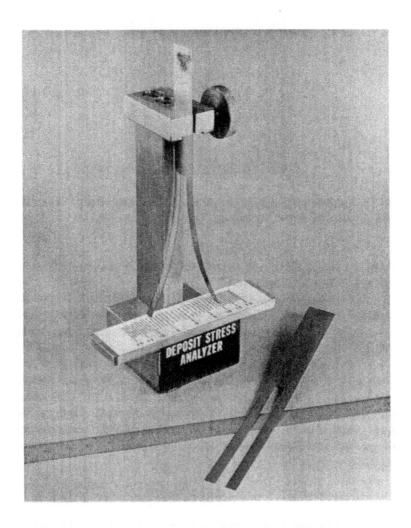

Figure 8: Another version of the flexible strip method for measuring stress.

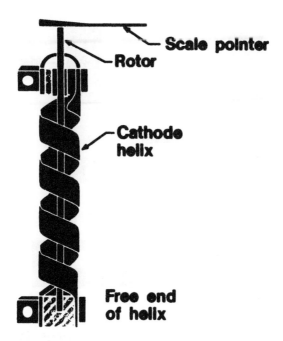

Figure 9:   Brenner and Senderoff's spiral contractometer for measuring stress. From reference 30.

Stresometer

The stresometer (Figure 10) combines Mill's thermometer bulb (31) with the bent strip method. The deposit is applied on one side of a thin metal disc. Beneath the disc but out of contact with the plating solution is a metering fluid connected to a precision capillary tube. When a stress develops in the deposit, the height of the liquid in the capillary tube changes. A tensile stress causes the disc to "dish in" while compressive stress causes the disc to bulge out. The rise or fall of the fluid is a direct linear measure of the stress in the deposit (32).

X-Ray

Although X-ray diffraction is widely used as a nondestructive method for the determination of macrostresses, it has found relatively little application for electrodeposits (24). The method is based on the changes in spacing between crystal planes associated with macrostresses. The problem is that it is difficult to determine the involved Bragg angles with sufficient accuracy because the diffraction lines are broadened, generally because of fine grain size and microstresses.

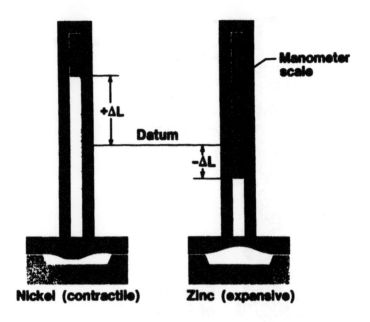

Figure 10: Kushner's stresometer for measuring stress. From reference 32.

Strain Gage

This technique provides for real time control of stress and has been used in the electroforming of optical components (33). Plating is done on a strain gage simultaneously with the part (Figure 11). As the plated surface of the gage bends in response to compressive or tensile forces, an analog output is produced. The strain signals are analyzed by computer programs which vary the output of the power supply up or down in response to compressive or tensile bending of the plated surface. Stress control with this method is reported to have been held sufficiently close to zero so that dimensional accuracy in optical nickel electroforms was 0.15 μm (6 millionths) (33).

Dilatometer

This method relies on the elastic expansion or contraction of a prestressed steel strip brought about by the force developed along its axis by the tensile or compressive stress in the deposit applied on its two surfaces (Figure 12). It offers the advantage of a continuous determination without some of the usual theoretical and practical drawbacks and gives results which compare well with those obtained by the rigid strip technique (34).

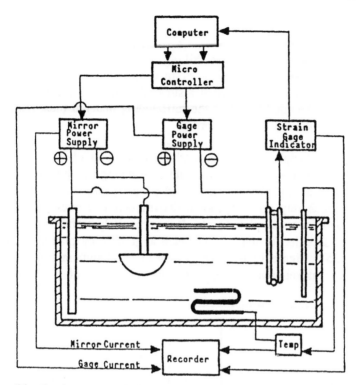

Figure 11: Strain gage technique for measuring stress. From reference 33.

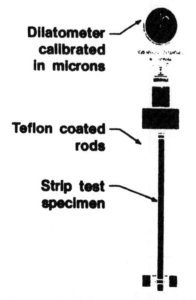

Figure 12: Dilatometer method for measuring stress. From reference 34.

Hole Drilling

This technique involves drilling a hole in the finished part and measuring the resulting change of strain in the vicinity of the hole. The method is based on the fact that if a stressed material is removed from its surroundings, the equilibrium of the surrounding material must readjust its stress state to attain a new equilibrium.

The principle is used quantitatively by drilling a hole incrementally in the center of strain-gage rosettes and then noting the incremental strain readjustments around the hole measured by the gages. Unlike most other methods which rely on independent determination of stress, this method provides data for actual plated parts. It is particularly useful for parts plated with thick deposits for applications such as joining by plating or electro-forming (35). Figure 13 is an edge view of an aluminum cylinder plated with thick nickel-cobalt alloy showing locations of residual stress

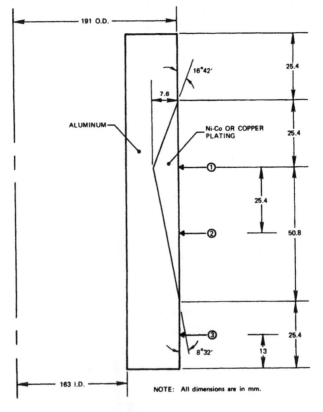

Figure 13: Edge view of plated aluminum cylinder showing locations of residual stress determinations. From reference 35.

determinations while Figure 14 shows maximum principal residual stress versus depth for various sections of the part.

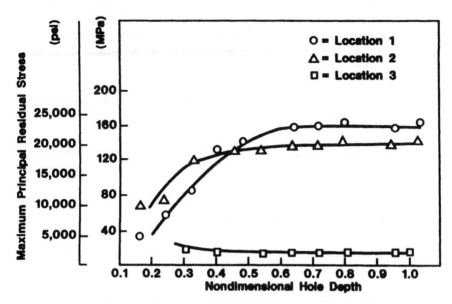

Figure 14: Maximum principal residual stress vs depth for an aluminum substrate and for two locations plated with Ni-40 Co alloy. See Figure 13 for hole locations.

Holographic Interferometry

Stored beam laser holography has been used to monitor stress in thin films during electrodeposition (36). The value of this technique is that it can be applied in situ and is applicable to very thin films, e.g., less than 20 μm. All the other techniques discussed in this chapter are typically used for thicker electrodeposits. This technique has practical advantages over conventional interferometry: it is similarly highly sensitive, can reveal the distribution of stress, and may be applied to diffusely reflecting surfaces.

STRESS THEORIES

Although a number of theories have been proposed to explain the origins of stress in electrodeposits, no overall theory that encompasses all situations has been formulated to date. Buckel (37) theorized that apart from thermal stress there may be as many as six other stress-producing mechanisms: incorporation of atoms (e.g., residual gases) or chemical reactions, differences of the lattice spacing of the substrate and the film during epitaxial growth, variation of the interatomic spacing with the crystal size, recrystallization processes, microscopic voids and dislocations, and phase transformations. The excellent review articles by Weil (24) detail the more prominent theories and the following information is excerpted from his publications. According to Weil, stress theories can be broken down into five categories:

1.   Crystalline material growing outward from several nuclei is pulled together upon meeting.

2.   Hydrogen is incorporated in the deposit, and a volume change is assumed when hydrogen leaves.

3.   Foreign species enter the deposit and undergo some   alterations thereby causing a volume change.

4.   The excess energy theory assumes that the overpotential is the cause of stress.

5.   Lattice defects, particularly dislocations and vacancies, are the cause of stress.

Crystallite Joining

This theory proposes that crystalline material growing outward from several nuclei is pulled together upon meeting. For example, the observation of vapor deposition in the electron microscope led to the discovery of the so-called liquid like behavior. Nuclei growing laterally were seen to act very similarly to two drops of liquid joining, that is, when they touched they immediately formed a larger crystal with a shape leaving a minimum surface area. This theory could explain certain conditions where three dimensional crystallites are nucleated. The resulting volume decrease to a more dense state would produce a tensile state, however, this theory does not explain how compressive stresses develop.

Hydrogen

This assumes that hydrogen is incorporated during deposition and a volume change occurs when the hydrogen leaves. For example, a layer of deposit contains hydrogen which may form a hydride with the metal species. Subsequently, the hydrogen diffuses out after the hydride, if formed, has decomposed causing a decrease in volume. The substrate and deposit beneath the layer which do not want to contract cause the tensile stress. Comnpressive stresses result if the hydrogen, instead of leaving the deposit, diffuses to favored sites and forms gas pockets.

Changes in Foreign Substances

Alterations in the chemical composition, shape, or orientation of codeposited foreign material have been postulated to cause the volume change of a plated layer, which originally fitted the one beneath it. This theory is based on very scant experimental evidence and more data are needed to support it.

Excess Energy

A metal ion in solution must surmount an energy barrier to be transformed from a hydrated ion to a metal ion firmly attached to the lattice. This may be thought of as a metal deposition overvoltage. Once the metal ion is over the hurdle, however, it possesses considerable excess energy; a group of such ions will have a higher temperature than their surroundings. The cooling down results in stress. This theory does not explain why compressive stresses are produced. According to Weil (24) this theory is simply a corollary to such other theories as those dealing with crystallite joining and dislocations.

Lattice Defects

The mechanical behavior of metals is now known to be determined primarily by lattice defects called dislocations. Most of the recently developed theories about the origins of internal stresses in deposited metals have included aspects of dislocation theory or are totally based upon it. Of these theories, the best developed is one which explains the misfit stresses between a deposit and a substrate of a different metal when the former continues the structure of the latter (Figure 15). Explanations of the intrinsic stresses in terms of dislocations have been developed theoretically, but there are not sufficient experimental data to verify them. In spite of this Weil (24) suggests that by a process of elimination, in many instances, the other theories do not apply, while the dislocation theory can at least explain the observed phenomena in a logical way.

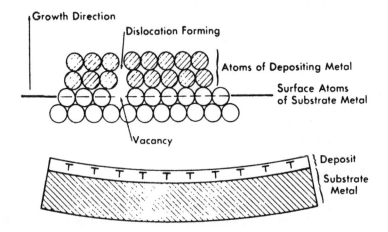

Figure 15: (Top) Edge dislocation forming in an electrodeposited metal near a surface vacancy in the basis metal. (Bottom) How an array of negative dislocations produces tensile stress in electrodeposits. From reference 2. Reprinted with permission of ASM International.

REFERENCES

1.   R.A. Collacott, "Residual Stresses", *Chartered Mechanical Engineer*, 26 (8), 45 (Sept 1979).

2.   J.B. Kushner, "Stress in Electroplated Metals", *Metal Progress*, 81, 88 (Feb 1962).

3.   M. Wong, "Residual Stress Measurement on Chromium Films by X-ray Diffraction", *Thin Solid Films*, 53, 65 (1978).

4.   S. Senderoff, "The Physical Properties of Electrodeposits Their Determination and Significance", *Metal Finishing*, 46, 55 (August 1948).

5.   K. Parker, "Effects of Heat Treatment on the Properties of Electroless Nickel Deposits", *Plating and Surface Finishing*, 68, 71 (Dec 1981).

6.   S.S. Tulsi, "Properties of Electroless Nickel", *Trans. Inst. Metal Finishing*, 64, 73 (1986).

7.   R. Rolff, "Significance of Ductility and New Methods of Measuring the Same", *Testing of Metallic and Inorganic Coatings, ASTM STP 947*, W.B. Harding and G.A. DiBari, Eds., American Society for Testing and Materials, Phil., PA, 19 (1987).

8.   A.T. Vagramyan and Z.A. Solov'eva, *Technology of Electrodeposition*, Robert Draper Ltd., Teddington, England (1961)

9.   W.H. Safranek, *The Properties of Electrodeposited Metals and Alloys*, Second Edition, American Electroplaters and Surface Finishers Society, Orlando, FL (1986).

10.   H.J. Noble and E.C. Reed, "The Influence of Residual Stress in Nickel and Chromium Plates on Fatigue", *Experimental Mechanics*, 14 (11), 463 (1974).

11.   J.E. Stareck, E.J. Seyb and A.C. Tulumello, "The Effect of Chromium Deposits on the Fatigue Strength of Hardened Steel", *Plating* 42, 1395 (1955).

12.   R.A.F. Hammond, "Stress in Hard Chromium and Heavy Nickel Deposits and Their Influence on the Fatigue Strength of the Basis Metal", *Metal Finishing Journal*, 7, 441 (1961).

13. N.P. Fedot'ev, "Physical and Mechanical Properties of Electrodeposited Metals", *Plating* 53, 309 (1966).

14. C. Williams and R.A.F. Hammond, "The Effect of Chromium Plating on the Fatigue Strength of Steel", *Trans. Inst. Metal Finishing*, 32, 85 (1955).

15. K. Lin, R. Weil and K. Desai, "Effects of Current Density, Pulse Plating and Additives on the Initial Stage of Gold Deposition", *J. Electrochemical Soc.*, 133, 690 (1986).

16. K. Parker and H. Shah, "Residual Stresses in Electroless Nickel, Plating", *Plating*, 38, 230 (1971).

17. J.B. Kushner, "Factors Affecting Residual Stress in Electrodeposited Metals, *Metal Finishing*, 56, 56 (June 1958).

18. J.L. Marti, "The Effect of Some Variables Upon Internal Stress of Nickel as Deposited From Sulfamate Electrolytes", *Plating 53*, 61 (1966).

19. A.F. Greene, "Anodic Oxidation Products in Nickel Sulfamate Solutions", *Plating* 55, 594 (1968).

20. J.W. Dini, H.R. Johnson and H.J. Saxton, "Influence of Sulfur on the Properties of Electrodeposited Nickel", *J. Vac. Sci. Technol.* 12, 766 (1975).

21. J.W. Dini and H.R. Johnson, "Electroforming of a Throat Nozzle for a Combustion Facility", *Plating and Surface Finishing*, 64, 44 (August 1977).

22. J.B. Kushner, "Factors Affecting Residual Stress in Electrodeposited Metals", *Metal Finishing*, 56, 81 (May 1958).

23. R. Weil, "The Measurement of Internal Stresses in Electrodeposits", *Properties of Electrodeposits, Their Measurement and Significance*, R. Sard, H. Leidheiser, Jr., and F. Ogburn, Eds, The Electrochemical Soc., Pennington, NJ, Chapter 19 (1975).

24. R. Weil, "The Origins of Stress in Electrodeposits", *Plating* 57, 1231 (1970), 58, 50 (1971) and 58, 137 (1971).

25. E. Raub and K. Muller, *Fundamentals of Metal Deposition*, Elsevier Publishing Co., New York (1967).

26.   L.C. Borchert, "Investigation of Methods for the Measurement of Stress in Electrodeposits", *50th Annual Technical Proceedings*, American Electroplaters Soc., 44 (1963).

27.   G.G. Stoney, "The Tension of Metallic Films Deposited by Electrolysis", *Proceedings Royal Society*, A82, 172 (1909).

28.   F.J. Schmidt, "Measurement and Control of Electrodeposition", *Plating* 56, 395 (1969).

29.   W.C. Cowden, T.G. Beat, T.A. Wash and J.W. Dini, "Deposition of Adherent, Thick Copper Coatings on Glass", Proceedings of the Symposium on Metallized Plastics: Fundamental and Applied Aspects, *The Electrochemical Soc.*, Pennington, NJ (Oct 1988).

30.   A. Brenner and S. Senderoff, "A Spiral Contractometer for Measuring Stress in Electrodeposits", *J. Res. Natl. Bur. Std.*, 42, 89 (1949).

31.   E.J. Mills, "On Electrostriction", *Proc. Royal Society*, 26, 504 (1877).

32.   J.B. Kushner, "A New Instrument for Measuring Stress in Electrodeposits", *41st Annual Technical Proceedings, American Electroplaters Soc.*, 188 (1954).

33.   R.W. George et al., "Apparatus and Method for Controlling Plating Induced Stress in Electroforming and Electroplating Processes", *U.S. Patent 4,648,944* (March 1987).

34.   W.H. Cleghorn, K.S.A. Gnanasekaran and D.J. Hall, "Measurement of Internal Stress in Electrodeposits by a Dilatometric Method", *Metal Finishing Journal* 18, 92 (April 1972).

35.   J.W. Dini, G.A. Benedetti and H.R. Johnson, "Residual Stresses in Thick Electrodeposits of a Nickel-Cobalt Alloy" *Experimental Mechanics*, 16, 56 (Feb 1976).

36.   F.R. Begh, B. Scott, J.P.G. Farr, H. John, C.A. Loong and J.M. Keen, "The Measurement of Stress on Electrodeposited Silver by Holographic Interferometry", *J. of the Less Common Metals*, 43, 243 (1975).

37.   W. Buckel, "Internal Stresses", *J. Vac. Sci. Technol.*, 6, 606 (1970).

10

CORROSION

INTRODUCTION

Corrosion (environmental degradation) is the destruction or deterioration of a material by chemical or electrochemical reaction with its environment. The National Materials Advisory Board in 1986 published a list of the ten most critical issues in materials and every single issue involved problems associated with corrosion. It was estimated that in 1985, corrosion problems cost the US over $160 billion as well as a countless number of lives (1). Corrosion is classified in a number of ways and a breakdown is shown in Figure 1 (2). A variety of factors including metallurgical, electrochemical, physical chemistry, and thermodynamic affect the corrosion resistance of a metal (Figure 2),and these all are part of the broad field of materials science(3). For that matter, one of the reasons this chapter is shorter than most of the others is the fact that corrosion is discussed in many of the other chapters. In many instances it is difficult to separate corrosion from many of the other property issues associated with deposits. For example, the tensile strength of a corroded sample can be reduced considerably as shown in Figure 3 because the cross sectional area is reduced by corrosion and therefore higher stresses are involved. In addition, the localized corrosion which has resulted in pits acts as stress raisers and deformation occurs prematurely at the pitted area (4). For more detail on corrosion, references 3-8 are recommended.

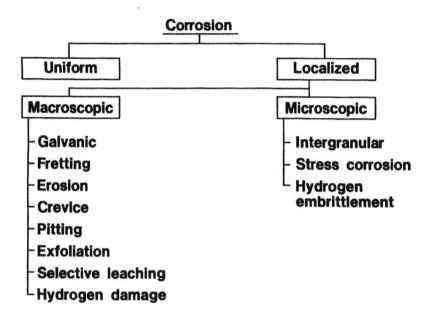

Figure 1: Corrosion classification. Adapted from reference 2.

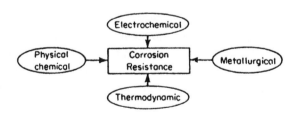

Figure 2: Factors affecting corrosion resistance of a metal. Adapted from reference 3.

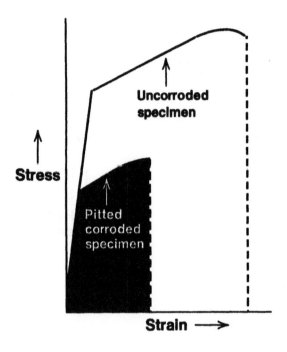

Figure 3: Scenario showing how corrosion can affect the tensile strength of a steel specimen. Adapted from reference 4.

SUBSTRATES

Engineering designs usually involve commercial alloys in various aqueous environments. The galvanic series in seawater (Table 1) is a useful guide in predicting the relative behavior of adjacent material in marine applications. Metals grouped together in the galvanic series have no appreciable tendency to produce corrosion, therefore, are relatively safe to use in contact with each other. By contrast, coupling two metals from different groups and, particularly, at some distance from each other will result in accelerated galvanic attack of the less noble metal. There can be several galvanic series depending on the environment of concern (9). In selecting a coating it is important to know its position with respect to its substrate in the galvanic series applicable for the intended service condition. Selection of a coating as close as possible in potential to the substrate is a wise choice because few coatings are completely free of pores, cracks and other defects (10).

Another item to consider is the interfacial zone between the basis

Table 1*: Galvanic Series

Galvanic Series of Metals and Alloys
Corroded End (anodic, or least noble)

Magnesium
Magnesium alloys
Zinc
Aluminum 2S
Cadmium
Aluminum 17ST
Steel or Iron
Cast Iron
Chromium-iron (active)
Ni-Resist
18-8 Chromium-nickel-iron (active)
18-8-3 Chromium-nickel-molybdenum-iron (active)
Lead-tin solders
Lead
Tin
Nickel (active)
Inconel (active)
Brasses
Copper
Bronzes
Copper-nickel alloys
Monel
Silver solder
Nickel (passive)
Inconel (passive)
Chromium-iron (passive)
18-8 Chromium-nickel-iron (passive)
18-8-3 Chromium-nickel-molybdenum-iron (passive)
Silver
Graphite
Gold
Platinum

Protected End (cathodic, or most noble)

* From Reference 9.

metal and its protective coating (11). This zone, which has three parts, can be responsible for the success or failure of the finished part.

Zone 1   includes the outermost surface of the basis metal viewed as a "skin". The significant thickness of this zone may vary from a few Angstroms to as much as 0.010 inch.

Zone 2   includes the first layer of the coating and may involve thicknesses of a few Angstroms to as much as 0.002 inch.

Zone 3   includes the alloy formed by diffusion of the coating and basis metal. Thickness may vary from a few Angstroms to 0.02 inch or more.

This three part interfacial zone and the metallic coating comprise a subject which has been referred to as Surface Metallurgy by Faust (11). The contribution of Zone I to overall performance is intimately tied in with the history of the basis metal and the kind of operations seen by its surface. Substrate metals that have been heavily worked by such operations as deep drawing, swaging, polishing and buffing, grinding, machining, forging and die drawing often come to the plating shop with a damaged layer on the surface that differs from the basis metal in grain size, structure and orientation (10). An example is shown in Figure 4. This heavily worked layer is termed a Beilby layer (12) and was originally thought to be amorphous or vitreous rather than crystalline. This weak, somewhat brittle layer was originally compared to the glass like form assumed by silicates when they are solidified from the molten state.   Further analysis has revealed that polishing occurs primarily by a cutting mechanism and that a Beilby layer is not formed (13). The polished surface is always crystalline, but is deformed and is inherently low in ductility and fatigue strength and therefore a weak foundation for plated coatings. For example, mechanically polished surfaces on stainless steel contain extremely fine grains in the form of broken fragments or flowed metal. Nickel deposits on this substrate are extremely fine-grained and bear no crystal relationship to the true structure of the basis metal. By comparison, use of electropolishing prior to nickel plating on stainless steel results in undistorted grains of normal size on which nickel builds pseudomorphically (14).

COATINGS

Metallic coatings are one method of preventing corrosion. Deposits applied by electrodeposition or electroless plating protect substrate metals in three ways: 1) cathodic protection, 2) barrier action, and 3) environmental modification or control (10). Cathodic protection is provided by sacrificial

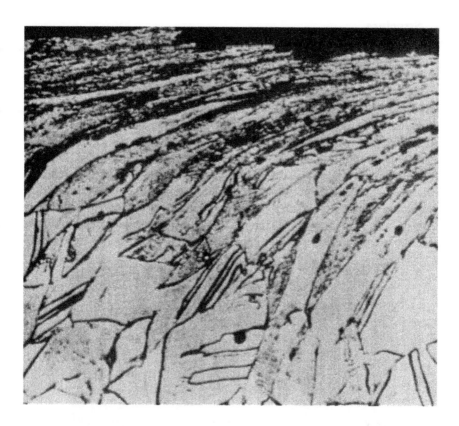

Figure 4: Cross section of a buffed metal surface showing severe distortion (200X). From reference 11. Reprinted with permission of ASM International.

corrosion of the coating, e.g., cadmium and zinc coatings on steel. Barrier action involves use of a more corrosion resistant deposit between the environment and the substrate to be protected. Examples of this include zinc alloy automotive parts and copper-nickel-chromium and nickel-chromium systems over steel (discussed in more detail later in this chapter). An example of environmental modification or control coatings in combination with a nonimpervious barrier layer is electrolytic tinplate used in food packaging (10).

Corrosion is affected by a variety of issues associated with coatings. These include structure, grain size, porosity, metallic impurity content, interactions involving metallic underplates and cleanliness or freedom from processing contaminants (15).

A. Structure

An example of the influence of coating structure in protecting a substrate from corrosion is aluminum ion plated uranium which shows significantly greater protection in a water vapor corrosion test with a dense noncolumnar structure than with a columnar structure. Figures 5 and 6 are aluminum ion plated coatings showing a structure that is columnar with large voids between columns (Figure 5) and a structure that is completely noncolumnar with no evidence of voids in the coating (Figure 6). Results of corrosion testing samples with these different structures are presented in Figure 7. The corrosion curve for coatings with the columnar structure similar to Figure 5 reveals only a minimum of protection with an incubation time of about 8 hours. By contrast, the corrosion test results for the noncolumnar structure shown in Figure 6 exhibit an incubation time on the

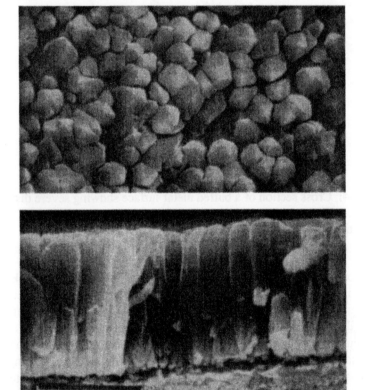

Figure 5: A columnar aluminum ion plated coating on uranium. From reference 16. The top view shows the surface morphology of the coating, while the bottom view shows a cross section. Reprinted with permission of the American Vacuum Society.

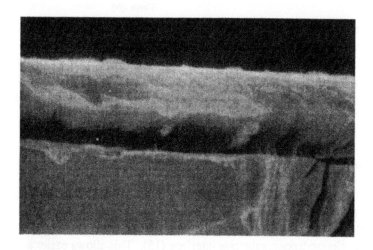

Figure 6: A noncolumnar aluminum ion plated coating on uranium. From reference 16. The top view shows the surface morphology of the coating, while the bottom view shows a cross section. Reprinted with permission of the American Vacuum Society.

order of 50 hours with a slower transition to linear corrosion kinetics that was not complete when the corrosion test was stopped (16).

Factors that favor nonepitaxial growth can cause gas porosity and voids to form at the interface between the substrate and deposit. For example, in the case of a copper substrate, if an acid dip is too strong so

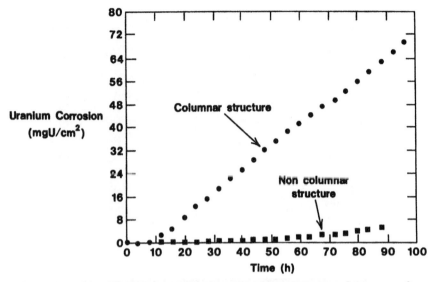

Figure 7: Corrosion of aluminum ion plated uranium samples exposed to a water vapor atmosphere. Adapted from reference 16.

that the etching results in development of large areas with {111} planes constituting the surface, the subsequently deposited films may not grow non-epitaxially) but also lose adhesion to the substrate forming an interfacial crack because of the voids (17).

B. Grain Size

Electrodeposits of small grain size, i.e., with a high area fraction of grain boundaries or those with columnar structure as discussed previously, exhibit a higher rate of transport of material between the external surface and the electrodeposit/substrate interface (15). This allows easier access for the corrosive species to the coating/ substrate interface. In addition, the substrate metal can diffuse more readily to the external surface to react with the environment.   This is particularly true at temperatures below 200 C (15), and is discussed in more detail in the chapter on Diffusion. Grain boundaries in a deposit tend to corrode preferentially and if there is a range of grain sizes, the fine-grained region tends to corrode (18). The crevices in fine-grained deposits also corrode preferentially. This is due to the fact that the grains in the crevices are even smaller than in the rest of the deposit and they also have a different chemical composition because of the greater incorporation of addition agent products (19).   It's also important to remember that stressed metal is anodic to annealed or lesser stressed metals and is therefore more prone to corrosion in unfavorable service conditions.

C. Porosity

Porosity in electrodeposits is such an important topic that an entire chapter is devoted to it in this book so only a few words will be said here. With sacrificial coatings such as zinc or cadmium on steel, porosity is not usually a problem since the coatings cathodically protect the substrate at the bottom of an adjacent pore. However, with noble coatings such as those used in electronic applications, substrates are subject to corrosion at pore sites. Porosity also permits the formation of tarnish films and corrosion products on surfaces, even at room temperature.

D. Codeposited Metallic Impurities

Codeposited metallic impurities can noticeably influence corrosion performance. For example, small amounts of sulfur in bright nickel deposits noticeably change the corrosion potential. This is discussed in detail in this chapter in the subsequent section on decorative nickel-chromium coatings. Small amounts of copper in bright nickel plating solutions cause significant reductions in salt spray resistance, e.g., 10 ppm of copper results in a 20% reduction, and 25 ppm of copper a 50% reduction (20)

E. Metallic Underplates

As mentioned throughout this book, underplates are used for a variety of purposes such as improving adhesion of the plated system, as diffusion barriers, or for improving mechanical properties. They are also important in improving corrosion resistance. One example is the use of a nickel layer between a copper substrate and a final gold deposit to prevent diffusion of the copper to the surface where it would subsequently tarnish (15).

F. Process Residues

The surface of a substrate must be free from soil and oxides before being plated. Besides assuring good adhesion this also prevents contaminants from being trapped at the interface and subsequently causing corrosion problems. Plating salts on the surface or in the pores of an electrodeposit also need to be adequately removed since they can increase the conductivity of adsorbed water and increase the probability of electrolytic or galvanic corrosion (15).

DECORATIVE NICKEL-CHROMIUM COATINGS

The progress made over the years in improving the corrosion resistance of decorative electrodeposited nickel-chromium coatings (with or without a copper underlayer) for automotive industry usage is a good example of a materials science and electrochemical study of the factors which influence the corrosion behavior of electrodeposits.   Figure 8 outlines the history of the development of nickel-chromium coatings (21). Significant highlights along the way included: 1) the understanding that thickness and composition of the nickel layer determined the corrosion resistance, 2) leveling copper and nickel plating processes, 3) semibright and bright nickel plating, 4) crack-free and microcracked chromium and 5) duplex and triplex (multilayer) nickel-chromium systems (10, 21).

The original nickel deposits were matte-gray and were polished after plating to produce a bright surface (21). Although these deposits were pore free and protected the underlying metal from corrosion, they had a tendency to tarnish in outdoor service. The development of the chromium plating process in 1924 was instrumental in eliminating this defect. Bright nickel-chromium coatings eventually replaced buffed nickel plus chromium. As the years went on and roads became more   salt-laden, bright nickel-chromium coatings were found to be unacceptable for the severe service conditions seen by automotive hardware exposed to these roads. This led to development of multilayer nickel coatings which effectively retarded the rate at which pits penetrated the coating by an electrochemical mechanism similar to cathodic protection. Bright nickel (0.04-0.15% sulfur) displays a more active dissolution potential (Figure 9) than semi-bright nickel containing about, 0.005% sulfur (22).   If the two deposits are electrically connected, the rate of corrosion of the bright, nickel is increased, whereas the rate of corrosion of the semibright nickel is decreased (23). By combining layers of nickel of different reactivity, lateral corrosion of the more reactive layer is enhanced thereby retarding penetration through this layer as shown in figure 10.

The corrosion resistance of a chromium deposit depends not so much on its thickness as on its physical state. If the chromium is crack-free or nonporous the corrosion resistance is excellent. However, chromium deposits typically do not remain crack-free in service. A small number of cracks are detrimental, however, the presence of many fine microcracks may be beneficial (10). This is due to the fact that microcracked chromium deposits cause the galvanic corrosion action to be spread over a very wide area.   Therefore, localized corrosion is avoided.   The influence of microdiscontinuities on the rate of pitting of nickel coatings is shown in figure 11. As the defect density of the chromium increases and pit depths and radii decrease; the average rate of pitting decreases from 1700 $\mu A/cm^2$

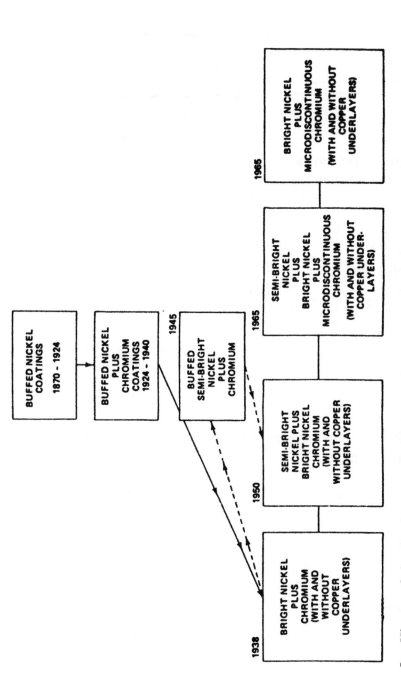

Figure 8: History of the development of nickel and nickel-chromium coatings. From reference 21. Reprinted with permission of Metal Finishing.

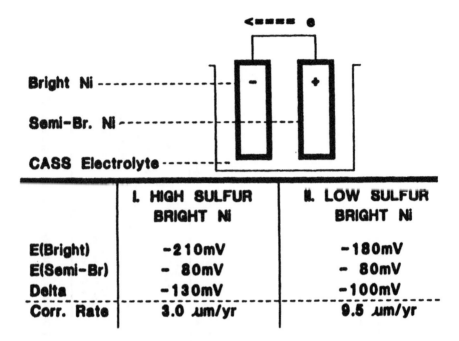

	I. HIGH SULFUR BRIGHT Ni	II. LOW SULFUR BRIGHT Ni
E(Bright)	-210mV	-180mV
E(Semi-Br)	- 80mV	- 80mV
Delta	-130mV	-100mV
Corr. Rate	3.0 μm/yr	9.5 μm/yr

Figure 9: Effects of sulfur content on corrosion in a duplex nickel composite. From reference 23. Reprinted with permission of The Electrochemical Society.

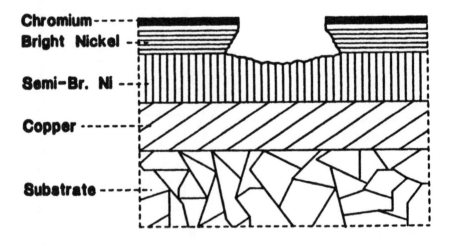

Figure 10: Corrosion scheme for corrosion site for a duplex-nickel system showing lateral penetration in the bright nickel layer. From reference 23. Reprinted with permission of The Electrochemical Society.

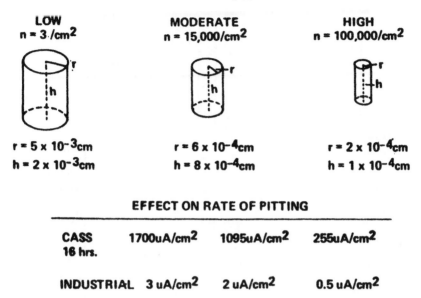

DEFECT DENSITY, n/cm²

LOW n = 3./cm²	MODERATE n = 15,000/cm²	HIGH n = 100,000/cm²
r = 5 x 10⁻³cm h = 2 x 10⁻³cm	r = 6 x 10⁻⁴cm h = 8 x 10⁻⁴cm	r = 2 x 10⁻⁴cm h = 1 x 10⁻⁴cm

EFFECT ON RATE OF PITTING

CASS 16 hrs.	1700uA/cm²	1095uA/cm²	255uA/cm²
INDUSTRIAL	3 uA/cm²	2 uA/cm²	0.5 uA/cm²

Figure 11: Effect of chromium pit characteristics and defect density on rate of pitting of nickel. From reference 21. Reprinted with permission of Metal Finishing.

to 255 µA/cm² in a 16 hour CASS test, while in an industrial environment, the rate of pitting decreases from 3 µA/cm² to 0.5 µA/cm². In the 16 hour CASS test, pit depths in nickel with conventional chromium varied from 10 to 20 µin while microdiscontinuous chromium pit depths were 1 to 8 µin (21). Figure 12 presents three year results for nickel-chromium and copper-nickel-chromium coatings on contoured steel panels in a marine atmosphere. Two types of nickel were included; Type I is the high-reactivity bright nickel and Type II the low reactivity bright nickel. The coating system with the high-reactivity double-layer nickel and microcracked chromium clearly provides the best overall performance (21).

CORROSION TESTS

A multitude of tests are available for evaluating the corrosion resistance of coatings but these will not be covered here. For detail see references 8,9, 24-26.

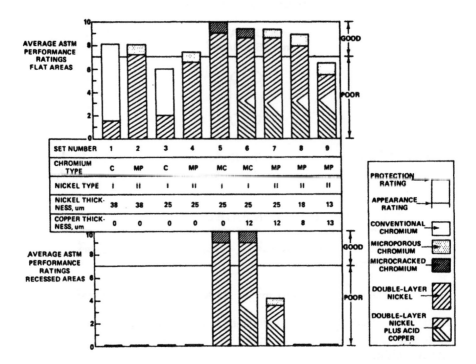

Figure 12: Performance of nickel-chromium and copper-nickel-chromium coatings on contoured steel panels in a marine atmosphere for 36 months. Ratings are for flat and recessed areas. From reference 21. Reprinted with permission of Metal Finishing.

REFERENCES

1.   R. Baboian, "Corrosion-A National Problem", *ASTM Standardization News*, 14, 34 (March 1986).

2.   E. Groshart, "Corrosion-Part 1", *Metal Finishing*, 84, 17 (May 1986).

3.   M. G. Fontana, *Corrosion Engineering*, Third Edition, McGraw-Hill (1986).

4.   H. McArthur, *Corrosion Prediction and Prevention in Motor Vehicles*, Ellis Horwood Ltd., England (1988).

5.   H. H. Uhlig, Editor, *The Corrosion Handbook*, Wiley & Sons (1948).

6.   P. A. Schweitzer, Editor, *Corrosion And Corrosion Protection Handbook*, Marcel Dekker, Inc., New York (1989).

7.   J. F. Shackelford, *Introduction to Materials Science for Engineers*, Second Edition, Macmillan Publishing Company, New York (1988).

8.   Metals Handbook, Ninth Edition, Volume 13, Corrosion, *ASM International*, Metals Park, Ohio (1987).

9.   L. Schlossberg, "Corrosion Theory and Accelerated Testing Procedures", *Metal Finishing*, 62, 57 (April 1964).

10.   J. Mazia and D. S. Lashmore, "Electroplated Coatings" in Metals Handbook, Ninth Edition, Volume 13, Corrosion, *ASM International*, Metals Park, Ohio (1987).

11.   C. L. Faust, "The Surface Metallurgy of Protective Coatings", *Metal Progress*, 71, 101 (May 1957).

12.   G. T. Beilby, "The Hard and Soft States in Metals", *J. Inst. of Metals*, 2, 149 (1912).

13.   L. E. Samuels, "The Nature of Mechanically Polished Surfaces", *46th Annual Technical Proceedings*, American Electroplaters Society (1959).

14.   C. L. Faust, "Smoothing by Electropolishing and Chemical Polishing", *37th Annual Proceedings*, American Electroplaters Society, 137 (1950).

15.   R. P. Frankenthal, "Corrosion in Electronic Applications", Chapter 9 in *Properties of Electrodeposits, Their Measurement and Significance*, R. Sard, H. Leidheiser, Jr., and F. Ogburn , Editors, The Electrochemical Soc. (1975).

16.   C. M. Egert and D. G. Scott, "A Study of Ion Plating Parameters, Coating Structure and Corrosion Protection for Aluminum Coatings on Uranium", *J. Vac. Sci. Technol.*, A5, 2724 (July/August 1987).

17.   E. C. Felder, S. Nakahara and R. Weil, "Effect of Substrate Surface Conditions on the Microstructure of Nickel Electrodeposits", *Thin Solid Films*, 84, 197 (1981).

18.   R. Weil and H. K. Tsourmas, "Early Stages of Corrosion of Nickel and Nickel-Chromium Electrodeposits", *Plating* 49, 624 (1962).

19.   R. Weil and W. N. Jacobus, Jr., "About Two Microstructural Features of Electrodeposits", *Plating* 53, 102 (1966).

20.   D. T. Ewing, R. J. Rominski and W. M. King, "Effect of Impurities and Purification of Electroplating Solutions. 1. Nickel Solutions (4), The Effects and Removal of Copper", *Plating* 37, 1157 (1950).

21.   G. A. DiBari, "Decorative Electrodeposited Nickel-Chromium Coatings", *Metal Finishing*, 75, 17, (June 1977) and 75, 17 (July 1977).

22.   *INCO Guide to Nickel Plating*, International Nickel Co., Saddle Brook, NJ (1989).

23.   J. H. Lindsay and D. D. Snyder, "Electrodeposition Technology in the Automotive Industry", in *Electrodeposition Technology, Theory and Practice*, L. T. Romankiw, and D. R. Turner, Editors, The Electrochemical Soc. (1987).

24.   *Corrosion Testing For Metal Finishing*, V. E. Carter, Editor, Butterworth Scientific, London, (1982).

25.   F. Altmayer, "Choosing an Accelerated Corrosion Test", *Metal Finishing*, 83, 57 (Oct 1985).

26.   *Testing Of Metallic And Inorganic Coatings*, W. B. Harding and G. A. DiBari, Editors, ASTM STP 947, American Society for Testing and Materials (1986).

11
WEAR

INTRODUCTION

Mechanical wear which is of great economical importance, accounting for losses of tens of billions of dollars a year, does not fit handily within the confines of a traditional discipline. Physics, chemistry, metallurgy and mechanical engineering are all of importance to the understanding of wear making it an ideal candidate to be included in a treatise on materials science (1). In fact, the study of wear is so complex that it can engage college students for an entire semester studying "tribology" which covers adhesion, friction and wear of materials.

Electrodeposits and their associated coatings such as electroless nickel, anodized aluminum and conversion coatings offer some distinguishing features for wear applications: a) since the coating processes are operated at less than 100°C, a hard surface can be applied to most metals and even plastics without regard for the metallurgical details of solid solubility and associated high temperature concerns that exist with most other processes, b) the low temperature aspects of these processes sometimes makes them the only coatings that can be applied to distortion prone substrates, c) since electrodeposits can be produced without causing distortion, they are often used for rebuilding worn parts, and d) the coatings can be applied in small holes and other recesses that are difficult to coat via other processes (2). The processes associated with plating and types of these coatings that find wide use are summarized in Figure 1. The coatings listed probably account for 90 percent of the plating family coatings used for wear applications (2).

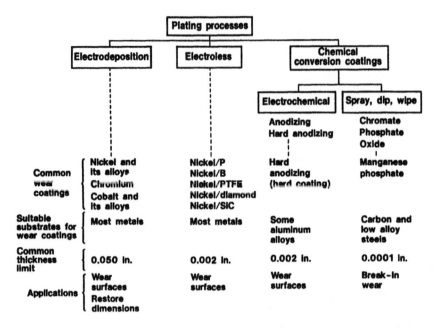

Figure 1:   Use of electrodeposition, electroless plating and chemical conversion coatings for wear applications. Adapted from reference 2.

This chapter will include some discussion on mechanisms of wear and testing of deposits for wear. Data for a variety of electrodeposits as well as electroless nickel and anodized aluminum will be presented as will some information on coatings for high temperature applications.

CLASSIFICATION OF WEAR

Wear is damage to a solid surface due to the relative motion between that surface and a contacting substrate or substances (1,2). It is a very complex phenomenon and usually involves progressive loss of material. Parameters that affect wear include:surface hardness and finish, microstructure and bulk properties, contact area and shape, type of motion, its velocity and duration, temperature, environment, type of lubrication and coefficient of friction. Budinski (2) has provided an excellent overview of wear processes, separating them into four categories based on commonality of mechanism as shown in Figure 2:

Abrasion — Wear produced when material is displaced through the action of a hard rough member on a softer surface by digging or grooving.

Erosion — Loss of material from a surface due to relative motion between a fluid and that surface.

Adhesion — Wear which is at least originated by the strong adhesion forces (akin to cold welding) between contacting members.

Surface fatigue — Removal of material as a result of cyclic stresses produced by repeated sliding or rolling on a surface. This is the main mode of failure of ball bearings.

Budinski (2) lists the following six traditional techniques for dealing with materials involved with wear processes:

- Use a lubricating film to separate conforming surfaces
- Insure that the wearing surface is hard
- Insure that the wearing surface is resistant to fracture
- Insure that the eroding surface is resistant to corrosion
- Insure that material couples are resistant to interaction in sliding
- Insure that the wearing surface is fatigue resistant

Coatings, regardless of the method of application, can offer help in most of the above techniques for dealing with wear.

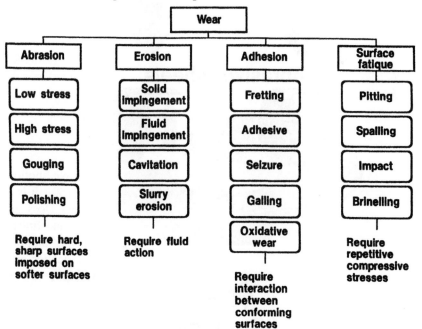

Figure 2: Basic categories of wear and modes of wear. From reference 2. Reprinted with permission of Prentice Hall, Englewood Cliffs, New Jersey.

WEAR TEST METHODS

There are numerous tests for evaluating wear resistance but it is not the purpose of this chapter to attempt to cover them all; those interested in more detail should consult references 2-6. However, some of the tests used to generate data presented in this chapter will be discussed. The two most common methods for testing wear resistance are the Taber Abrader and Falex Lubricant Tester.

Taber Abrader (Figure 3)

This test is designed to evaluate the resistance of surfaces to rubbing abrasion. Wear is produced by the sliding action of two abrading wheels against a rotating test sample. As shown in Figure 3, one abrading wheel rubs the specimen outward toward the periphery and the other inward. The weight loss in mg per 1000 cycles is measured and expressed as the Taber Wear Index. The lower the Taber Wear Index, the better the wear

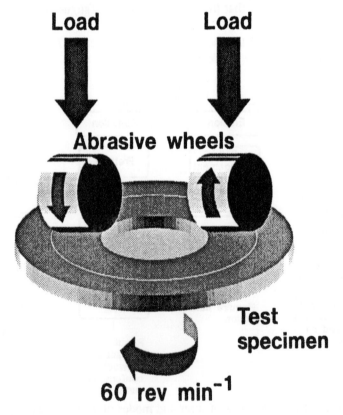

Figure 3: Taber abrader wear test.

resistance of the coating. The Taber Abrader uses no lubricant and determines wear under dry friction conditions. Rotating wheels of different hardness and composition are available. The Taber Abrader provides data for light abrasive wear conditions.

Falex Lubricant Tester (Figure 4)

This measures the wear of a 0.25 inch diameter pin rotating at 290 rpm between two stationary V-grooved blocks under variable load. The blocks are spring loaded in a jaw mechanism similar to a nutcracker. The total load is increased mechanically to any desired level up to 1800 kg by means of a rachet. The Falex Tester provides data on adhesive wear in lubricated conditions in a unidirectional action.

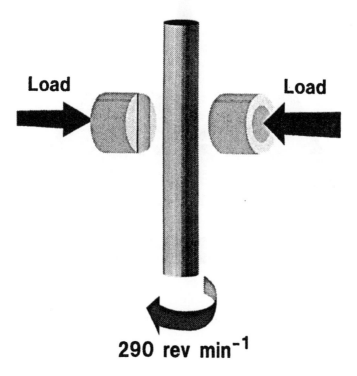

Figure 4: Falex wear test.

Reciprocating Scratch Test (Figure 5)

This test involves the tip of a Rockwell diamond indenter sliding under a 2 kg load against a flat coated specimen.

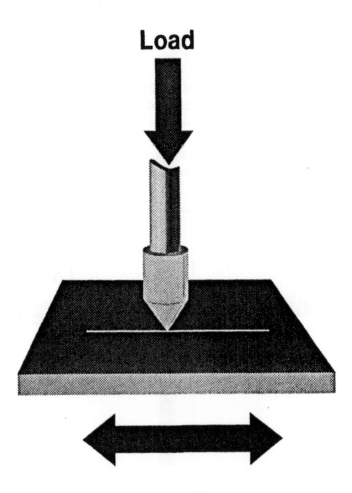

Figure 5: Reciprocating diamond scratch wear test.

Pin-on-Flat (Figure 6)

In the pin-on-flat test, the pin moves relative to a stationary flat in a reciprocating motion. The pin can be a ball, a hemispherically tipped pen, or a cylinder.

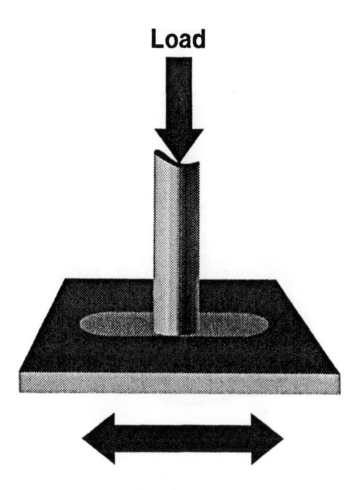

Figure 6: Pin-on-flat wear test.

Alfa Wear Test (Figure 7)

This test subjects samples to high pressure, adhesive wear under clean, lubricated conditions. A rectangular block is run against the periphery of a rotating hardened steel ring under known conditions of load, sliding velocity, and lubrication. The block is either a homogeneous wear resistant material or is made of steel and then coated with the wear resistant material to be tested (6).

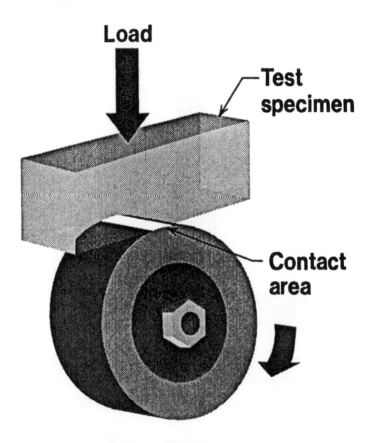

Figure 7: Alfa wear test.

Accelerated Yarnline Wear Test (Figure 8)

This test was designed to simulate typical conditions commonly found in textile machinery. A full-dull 1.5 mil diameter nylon monofilament is drawn at 1000 yards/min and 10 grams of tension through a layer of 1 micron aluminum oxide powder just inches before encountering the cylindrical test sample.

CHROMIUM

Chromium plating is more extensively used for wear applications than any other electrodeposited coating. Typical uses include roll surfaces, shaft sleeves, pistons, internal combustion engine components, hydraulic

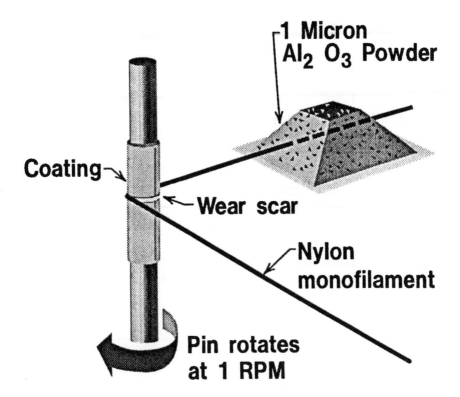

Figure 8: Accelerated yarnline wear test.

cylinders, landing gear and machine tools (7,8). Although the thickness varies with the application, it is usually in the range of 20 to 500 μm. By contrast, this is noticeably thicker than the 1 μm thick deposit referred to as decorative chromium. Although the term "hard chromium" has been used to describe the thicker deposit, there is no evidence that this deposit is any harder than decorative chromium (7).

Hard chromium plating exhibits better resistance to low stress abrasion than hard anodized aluminum and heat treated electroless nickel (Figure 9). It has a wear rate an order of magnitude lower than hard anodized aluminum, its closest competitor. By contrast, soft metals such as cadmium and silver perform poorly in terms of abrasion resistance (8).

Figure 10 presents reciprocating scratch wear data for conventional and crack-free chromium and electroless nickel coatings as a function of number of cycles (4,9). The results show that the conventional chromium coating with the highest hardness (as-deposited) exhibits the lowest wear rate. Heat treating the chromium deposit, which drastically affects its

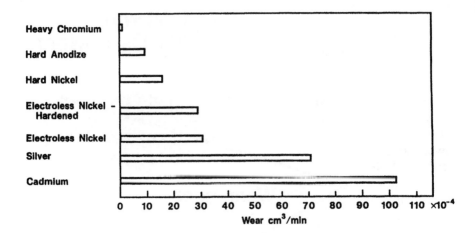

Figure 9: Abrasion rate of various coatings in the ASTM G 65 dry sand/rubber wheel test. Adapted from reference 8.

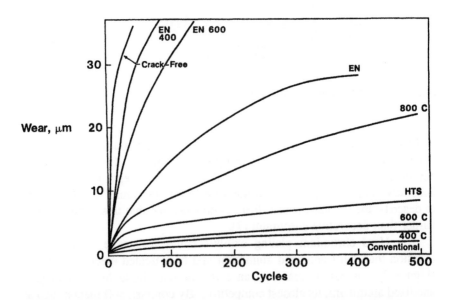

Figure 10: Wear of conventional as-deposited and heat treated (400, 600 and 800°C) chromium plating on a hard substrate and on a heat treated (softened) substrate (HTS), electroless nickel (EN), and heat treated electroless nickel (EN 400 and EN 600) in the reciprocating diamond scratch test. The electroless nickel contained 8.5 wgt % P. Adapted from references 4 and 9.

hardness, (e.g., one hour at 800°C reduces the hardness from 900 to 450 kg/mm²) results in increasingly higher wear rates with increasing temperature (4,9).

The most striking feature of Figure 10 is the very high wear rate of the crack-free chromium coating. This high wear rate is related to its crystal structure (4). Crack-free chromium has a predominantly hexagonal close-packed crystal structure unlike conventional chromium which is body-centered cubic. HCP metals tend to slip on only one family of slip planes, those parallel to the basal plane. This results in larger strains at a given stress level and less dislocation interactions. In addition, the strain-hardening rate is low, leading to rapid localization of deformation, early fracture and an increased wear rate (4).

Figure 11 shows results obtained with the Falex test. Once again, conventional chromium deposits show superiority when compared with electroless nickel and electrodeposited nickel coatings.

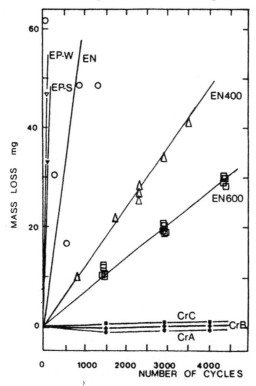

Figure 11: Wear of a variety of hard chromium deposits (Cr A, Cr B, Cr C), electroless nickel 8.5% P (EN), heat treated electroless nickel (EN 400 and EN 600), Watts electroplated nickel (EP-W) and sulfamate electroplated nickel (EP-S) in the Falex test. From reference 5. Reprinted by permission of the publisher, Elsevier Sequoia, The Netherlands.

CHROMIUM PLUS ION IMPLANTATION

Although electrodeposited chromium performs well in applications in which abrasion is severe or in which the wear mode is adhesive in nature, further treatment of the chromium can improve performance even more. An example is the use of ion implantation which is finding increased usage in enhancing wear, fatigue and corrosion resistance of metals. Ion implantation involves the injection of atoms into the near surface of a material at high speeds to form a thin surface alloy (10). No dimensional change occurs as a result of this process. Parts such as tools, dies, and molds exhibit longer life if the hard chromium deposit is followed by ion implantation with nitrogen. Electron diffraction studies have shown that the implanted layer is transformed to Cr_2N, resulting in an approximately 25% volume expansion of the lattice. According to some researchers, this volume increase closes the microcracks in the implanted region and significantly increases the load bearing capacity of the surface (11,12). More recently, Terashima et al., reported that although ion implantation with nitrogen resulted in the formation of Cr_2N, cracks in the chromium were not healed (13). Regardless, they also noted a remarkable improvement in wear resistance and improved corrosion resistance. Figure 12 shows results from pin on disc tests for unimplanted and nitrogen implanted chromium plated Ti-6Al-4V. A wear rate decrease of at least a factor of 20 was achieved at loads of 5.2 and 10.5 N when nitrogen implantation was used (11). Ion implantation also improves the corrosive part of the abrasive-corrosive wear process in certain applications (14). Practical examples of the use of ion implantation with chromium plated parts can be found in references 10 and 15.

ELECTROLESS NICKEL

The resistance of electroless nickel layers to wear is one of their remarkable properties. Some typical applications where these coatings are used to reduce wear include: hydraulic cylinders, pumps, valves, sliding contacts, shafts, connector pins, impellers, rotor blades, heat sinks, bearing journals, clutches, relays, drills, taps, and gears.

Although wear related properties of electroless nickel deposits are good, the recent development of low phosphorus electroless nickel coatings offers even further property enhancement (16). By way of definition,low P coatings contain 1-5% by weight P, medium P deposits 5-8% P,and high P deposits 9-12% P. Taber results presented in Figure 13 show that low phosphorus deposits have far superior abrasion resistance to alternate electroless nickel deposits and compare favorably with hard chromium and

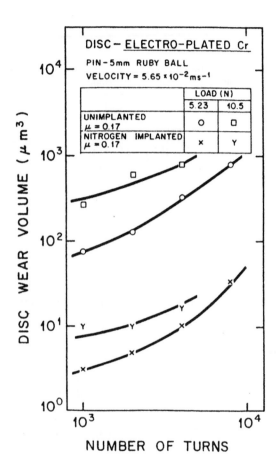

Figure 12:   Pin-on-disc wear data for unimplanted and nitrogen ion implanted electroplated chromium.   From reference 11.   Reprinted by permission of the publisher, ASM International, Metals Park, Ohio.

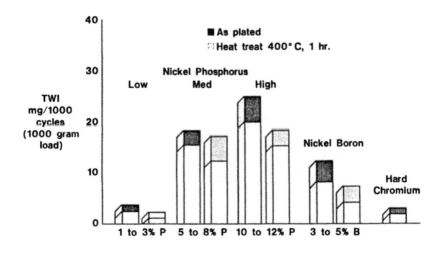

Figure 13: Taber abraser wear test results (CS-10 wheel) for several electroless nickel-phosphorus deposits. Adapted from reference 16.

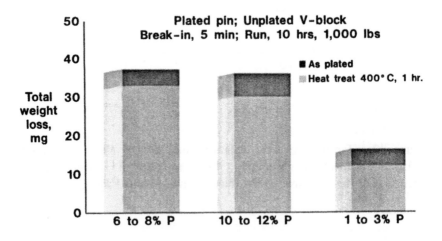

Figure 14: Falex wear test results for several electroless nickel-phosphorus deposits. Adapted from reference 16.

high boron nickel coatings (16). Low phosphorus deposits also show superior resistance to adhesive wear in Falex tests when compared with other electroless nickel deposits (Figure 14).

With medium P electroless nickel (8.5% P),there is no simple correlation between hardness and wear (17). Falex and pin-on-flat tests place electroless nickel in a different ranking order than that obtained with the diamond scratch test. As shown in Table 1, heat treatment reduced the wear rate of electroless nickel in all tests but the scratch test. This is due to the fact that the dominant wear mechanism changes from adhesive transfer to abrasive wear. This demonstration that the relative wear rates of materials depends on the type of wear test method emphasizes the importance of wear diagnosis in materials selection and design. An essential first step is the examination of worn components to identify the predominant wear mechanism (4,17).

Table 1 - Effect of test method on relative wear rate of chromium and electroless nickel deposits

Test Method	Relative wear rate (a)				
	Cr A600(b)	Cr D(c)	EN (d)	EN 400	EN 600
Reciprocating diamond scratch	2	71	14	46	33
Falex	-	--	165	32	19
Pin-on-flat	-	--	38	6.2	--
Taber	-	--	5.0	4.1	3.3

a.   This table is from reference 17. Relative wear rate equals wear rate of coating under specified test divided by wear rate of conventional chromium plating under same test. Chromium plating is used as the standard because its ranking order in terms of wear amongst the other coatings does not change with the test method.

b.   This is conventional chromium which has been heated at 600 C for 1 hour.

c.   This is crack free chromium

d.   The electroless nickel coatings contained 8.5%(wgt) P. EN 400 and EN 600 refer to one hour heating at 400 and 600 C, respectively.

ELECTROLESS NICKEL WITH DISPERSED PARTICLES

Inert particles are sometimes deposited with electroless nickel. Coatings of this type are often called composite coatings and although a later section in this chapter will discuss composite coatings, those involving electroless nickel will be covered here. The process involves the codeposition of diamond particles or powdered ceramics such as aluminum oxide and silicon carbide. The particles are suspended in stabilized electroless nickel-hypophosphite solution by mechanical or air agitation and randomly included during the formation of the coating. The particles can constitute up to 30 percent of the volume of the deposit and generally enhance hardness and wear resistance.

The particle coatings have a dull and rough appearance, but can be polished to a smooth, semi-bright finish. For most applications, the optimum particle size is in the range of 1 to 10 μm. Deposit thickness generally ranges from 10 to 35 μm for diverse applications such as metal forming dies, oil well tubes and molds for plastic materials that contain abrasive fillers. From the variety of particulate matter that can be codeposited, commercial attention has been focused primarily upon aluminum oxide, polycrystalline diamond, silicon carbide and PTFE (polytetrafluorethylene).

The superiority of polycrystalline diamond in an electroless nickel matrix is shown in Figure 15 which presents Alfa wear testing data for both test specimens (coating sample) and the contacting surface (5). Table 2 includes typical results from Taber testing, and based on these data, the

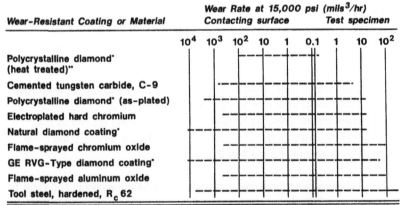

Figure 15: Alfa wear test results for various materials. Adapted from reference 5.

wear lifetime for the composite diamond coating is expected to be four times better than hard chromium plating. This has been verified by field testing (5). Others have also obtained excellent wear resistance with these types of coatings, particularly in the textile industry and for paper handling machines (6,18,19). However, diamond composite coatings are not well suited to resisting high pressure abrasive or adhesive wear. Contact pressures in excess of about 25,000 to 30,000 psi cause the diamond particles to be dislodged from the coating (6).

PTFE is a chemically inert, slippery polymer capable of continuous operation under cryogenic conditions or at temperatures up to 290°C. When an electroless nickel/PTFE composite surface suffers wear during usage, fresh PTFE is exposed to the wearing surface thereby ensuring a continuous supply of lubricant. Electroless nickel containing PTFE is not suitable for abrasive wear situations nor for applications involving high loads. However, under low loading, high cycle usage, its performance is excellent (20). Applications include carburetor components, butterfly valve discs, armature shafts in windshield washer pumps, lock components, and circuit breaker components. Friction test results comparing a PTFE/EN composite

Table 2 - Taber Wear Data for Various Coatings*

Wear Resistant Coating or Material	Per 1,000 cycles (10^4 mils^3)	Wear Rate Relative to diamond
Polycrystalline diamond**	1.159	1.00
Cemented tungsten carbide Grace C-9 (88 WC, 12 Co)	2.746	2.37
Electroplated hard chromium	4.699	4.05
Tool steel, hardened, R_c62	12.815	13.25

* From Reference 5
** Composite coating contained 20 to 30% of a
3-μm grade diamond in an electroless nickel matrix.

(20% volume PTFE and 5-10% P) with standard (5-9%)P and high (9-12%)P electroless nickel coatings at the same thickness of 0.4 mil (10 um) and under the same conditions are shown in Figure 16. The traces of the coatings without PTFE illustrate their classic galling behavior (21).

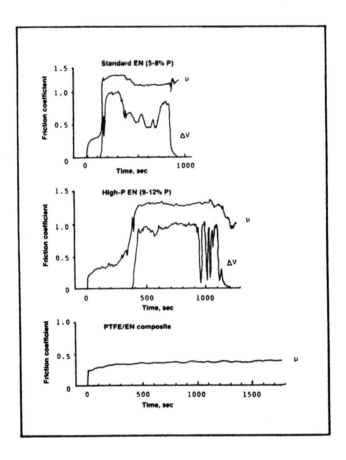

Figure 16: Comparison of 10 μm thick composite and electroless nickel coatings. Traces labeled u indicate friction coefficients. The other traces indicate changes in contact resistance caused by formation of wear debris. From reference 21. Reprinted by permission of the publisher, American Electroplaters & Surface Finishers Soc., Orlando, FL.

Almost no steady state wear occurred for either coating before the onset of abrasive wear. By comparison, the composite performed under a steady state regime up to 3800 seconds. The preferred range of PTFE is around 20% by volume as verified by Figure 17. Friction coefficients for coatings containing 9 or 15% by volume PTFE increased rapidly with time and even at 18% by volume, the trace illustrates the onset of abrasive wear at an early stage of testing (21).

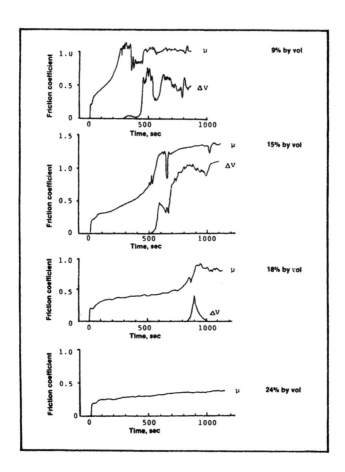

Figure 17: Effect of increasing PTFE content on wear resistance of composite coatings. Traces labeled u indicate friction coefficients. The other traces indicate changes in contact resistance caused by formation of wear debris. From reference 21. Reprinted by permission of the publisher, American Electroplaters & Surface Finishers Soc., Orlando, FL.

ELECTROLESS NICKEL PLUS CHROMIUM

The use of electroless nickel as an undercoat prior to hard chromium plating provides the advantages of both deposits. The hardness and wear resistance of chromium are retained while corrosion resistance is improved. Coverage of electroless nickel is uniform and not related to throwing power as is hard chromium, so the initial electroless nickel layer provides a uniform protective envelope (22,23). A variety of applications including aircraft, food industry, plastic molds and hydraulics attest to the viability of this coating combination. Taber wear data presented in Table 3 show that the presence of electroless nickel under chromium deposits provides slightly lower wear numbers than chromium by itself. This may be due to the thinner 0.7 mil (17.5 µm) layer of chromium having less nodulation and roughness than the thicker 2 mil (50 µm) layer used without an electroless nickel undercoat (23).

Table 3 - Taber wear test results for electroless nickel, chromium and electroless nickel plus chromium deposits (a)

Coating	Thickness (mil/µm)	Taber Wear Index
EN	2.0/50	16.3
EN	"	12.7
EN	"	13.6
Cr	"	0.8
Cr	"	1.0
Cr	"	0.9
EN/Cr	(.3/.7)(7.5/17.5)	0.9
EN/Cr	" "	0.5
EN/Cr	" "	0.6
EN/Cr	(.5/.5)(12.5/12.5)	0.5
EN/Cr	" "	0.5
EN/Cr	" "	0.5

a - These data are from reference 23

PRECIOUS METALS

A. Gold

The wear properties of gold deposits are quite important in many applications (24). Examples include current carrying devices such as electrical connectors, instrument slip rings and switches. Decorative applications such as jewelry, watch cases and table wear also require gold plated wear resistant surfaces. Thin gold coatings are used in aerospace bearing applications since gold shears easily and this is an important lubricant requirement (24).

Wear results shown in Figure 18 for a variety of gold deposits reveal that in most cases there is an inverse relationship between wear and hardness of the deposit (25). Deposits containing nickel, cobalt and cobalt-indium (hardnesses above 225 Knoop) exhibit the least wear while

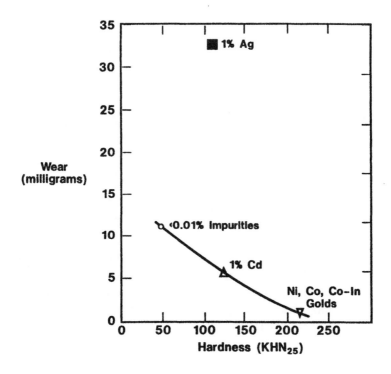

Figure 18: Wear test data for a variety of gold deposits. Adapted from reference 25.

pure gold with a hardness of less than 50 Knoop wears considerably more. An anomaly, however, is the result for gold containing one per cent silver which has the same hardness as gold containing one per cent cadmium yet exhibits more than an order of magnitude more wear. This is proof that one cannot always make the judgement that hardness is related to wear resistance (25).

With layered materials such as noble metal contacts made by electroplating or cladding, increasing substrate and underplate hardness may provide help in reducing wear, particularly if the coatings are thin (26). An example of the value of a hard underplate such as nickel in preventing adhesive wear is shown in Figure 19 which also reveals the superiority of a gold-cobalt electrodeposit compared to pure gold.

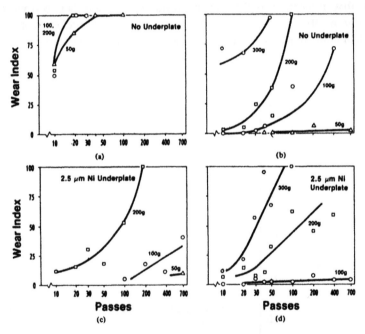

Figure 19: Electrographic wear indexes from unlubricated wear runs with 3.3 μm thick gold electrodeposits-(a), (c), ductile pure gold, with and without 2.5 μm nickel underplate and (b), (d), hard cobalt gold, with and without 2.5 μm, nickel underplate. From reference 26. Reprinted by permission of IEEE, Piscataway, NJ.

B. Palladium

One alternative to gold is palladium and its alloys of silver, nickel and cobalt, overplated with a thin 5 μm (0.125 μm) layer of hard gold. This combination has been used on production connector equipment for nearly ten years with substantial cost reductions (27). Electrodeposited palladium also offers other advantages: its hardness is in the range of 300-325 (KHN25) compared to 130-200 for hard gold, implying a more wear resistant surface, the toxicity of the ammoniacal or amine palladium solutions is much lower than the cyanide solutions used for hard gold deposits (27), and gold plated palladium looks very much like gold, making the finished product more appealing to the customer. The friction properties of electroplated palladium are far inferior to those of electroplated gold, however, electroplated palladium with a thin gold layer over it exhibits friction properties similar to those of conventionally available gold contacts. The gold layer over the palladium plays an effective role as a lubricant in maintaining friction properties (28). Figure 20 compares sliding contact test results for a variety of palladium and gold coatings over a nickel underlayer. A combination of 0.5 µm palladium plus 0.1 µm of gold performed about as well as 0.75 µm of gold. Also note in Figure 20 the large wear scars developed with wrought palladium and wrought gold after only one pass.

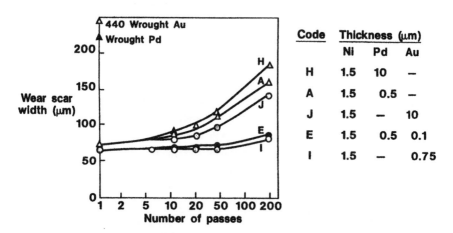

Figure 20: Relationship between wear scar width and number of passes for a variety of Ni, Pd, Au coatings plated on phosphor bronze wire. Adapted from reference 28.

C. Rhodium

The properties of rhodium are particularly well suited to many electrical and electronic applications. It has been used extensively in situations requiring wear resistance and stability at high temperatures. In general, rhodium improves efficiency whenever a low resistance, long wearing, oxide-free contact is required. Rhodium deposits assure low noise level for moving contacts, no oxide rectification and low and stable contact resistance (29).

D. Silver

Silver has excellent bearing properties, provides a surface resistant to galling at low loads, and will wet and retain an oil film. For unlubricated metal-to-metal wear, silver provides better wear characteristics than other coatings (Figure 21). In this situation chromium provided the closest wear rate to silver, but produced significant counterface wear (8).

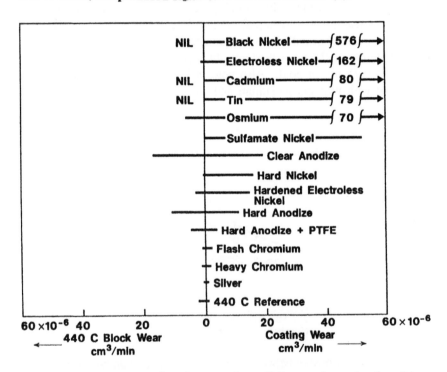

Figure 21: Wear rate of various coatings sliding against a hardened type 440C stainless steel counterface. Adapted from reference 8.

OTHER ELECTRODEPOSITED COATINGS

Nickel deposited in either a sulfamate solution or sulfate/chloride ("hard" nickel) solution is frequently used to rebuild worn or poorly machined parts. However, the hardness of 400 to 500 HV for the latter dictates grinding after plating whereas sulfamate nickel deposits can be machined. Electrodeposited tin-nickel (65Sn-35Ni) is used as an underplate for thin hard gold contact finishes and osmium deposits are sometimes used as wear counterfaces for jewels in watch improvements. Thin deposits of ruthenium (0.6 μm) have exhibited better wear properties than comparably thick rhodium films (30).

ANODIZED ALUMINUM COATINGS

A. Introduction

Anodized (oxide) coatings on aluminum offer good abrasion and corrosion resistance to the underlying metal. Conventional anodic coatings are employed for decorative or protective purposes while the thicker, more dense anodic films referred to as "hardcoating" or "hard anodic coating" are used for wear and abrasion resistance, since they perform much better in these applications as indicated in Table 4 (31). Some applications requiring hard anodized coatings for abrasion resistance include gears, nozzles, pinions, impeller blades and helicopter blades. Typical thickness of coating for optimum results is around 25 μm (1 mil). Some confusion occasionally arises regarding hardness of the coating. The term "hard anodic coating" is misleading since this coating is no harder than conventionally anodized

Table 4 - Abrasion Resistance of Conventional and Hard Anodic Coatings (a)

Number of Revolutions	Weight Loss (gm)	
	Conventional Anodic Coating	Hard Anodic Coating
1,200	0.0032	0.0017
2,200	0.0042	0.0019
4,200	0.0073	0.0033
10,000	0.0116	0.0058
15,000	0.0164	0.0081
20,000	0.0206	0.0103

a - This table is from reference 31. The data are from Taber abrasion tests using C517 discs with 100 gram loading on 6061 aluminum alloy. The coatings were approximately 42.5 μm (1.7 mils) thick.

aluminum. Both coatings are essentially aluminum oxide (31). However, due to the difference in operating conditions during anodizing, the hard coating process produces a denser oxide and this results in increased abrasion and erosion resistance (31,32). Excellent coverage on anodizing of aluminum and its properties can be found in the two volume text by Wernick et. al (33). Hard anodized coatings compare quite favorably with hard chromium plating in Taber wear tests as shown in Figure 22.

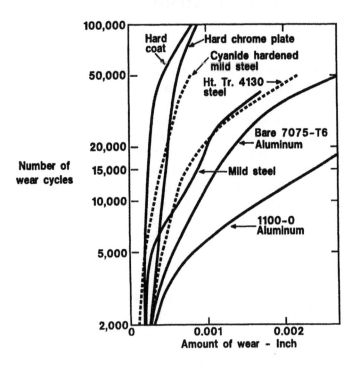

Figure 22: Taber abraser (CS-17 wheel, 1000 g load) results comparing hard anodized 7075-T6 aluminum with various other coatings and materials. Adapted from reference 31.

Alloy composition can significantly affect the properties of hard anodic coatings as shown in Figure 23. Taber abrasion tests using a CS-17 Taber wheel with a 500 gram load for 10,000 revolutions at 70 rpm revealed that hard coatings on 6061-T6 alloy perform significantly better than on 2024-T3 (34).

B. Surface Finish

The resistance to wear of an anodic coating is closely related to the surface finish (35). Therefore, it is important to recall that surfaces become

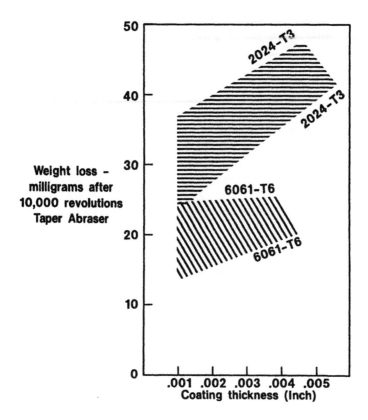

Figure 23: Taber abraser weight loss versus coating thickness for two different aluminum alloys. Adapted from reference 34.

slightly rougher during hard anodizing. Under normal anodizing conditions the natural roughness will increase by about 10-20 microinches (0.25-0.5 µm) for wrought alloys, while for casting alloys this increase in roughness may be between 50-100 microinches (Table 5).

C. Sealing

After anodizing, parts are rinsed thoroughly and may then be sealed in boiling distilled water, 5% dichromate solution, dewatering oil or wax (80°C, 15-30 minutes). Dichromate sealing improves the fatigue properties but, in common with other aqueous sealing solutions, decreases the abrasion resistance (Figure 24). For this reason, aqueous sealing processes are not normally used where high wear resistance is required (36).

**Table 5 - Surface Smoothness Before and
After Hard Coating (a)**

		Surface smoothness (µin. RMS)	
Aluminum alloy	Coating thickness (mil)	Initial	Final
7075-T6	1.2	5-7	10-20
2014-T6	1.3	5-8	50-60
356-T6	1.3	5-7	60-70
A214	1.1	6-10	40-45
43	1.1	8-10	60-70

a - This table is from reference 35

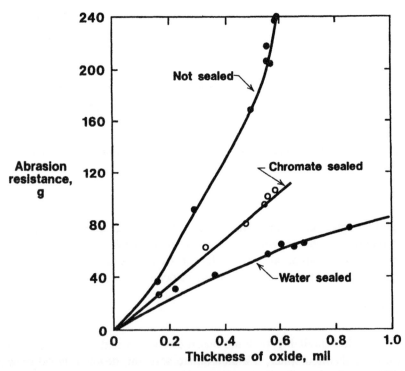

Figure 24:   Relation of abrasion resistance to thickness of oxide for different sealing conditions. Abrasion resistance was measured in grams of 180-mesh silicon carbide abrasive required to wear a 2 mm diameter area through the oxide film. Adapted from reference 36.

COATINGS FOR HIGH TEMPERATURE APPLICATIONS

Electrodeposited coatings are typically not effective for applications requiring wear resistance at temperatures above 500 to 600°C. However, use of composite coatings offers promise for higher temperatures. An electrodeposited composite coating consisting of a matrix of cobalt and containing approximately 30% by volume chromium carbide powder (2-4 μm diameter) is capable of controlling certain forms of wear on aircraft engines at temperatures up to 800°C and has been used on tens of thousands of parts (37). The wear resistance of the coating stems principally from the formation of a cobalt oxide glaze on the load bearing contact areas during interfacial motion. In similar fashion, a composite containing 20 to 25% percent by weight, of Cr_2O_3 in a cobalt matrix exhibited excellent resistance to adhesive wear at temperatures of 300 to 700°C due to the formation in air of an oriented oxide layer of Co_3O_4 (38). This electrodeposited composite was much better resisting wear at elevated temperatures than composite coatings of Ni/TiC or Ni-P/Si as shown in Figure 25. CoCrAlY overlay finishes have been successfully applied to turbine blades by electrodeposition of composite coatings followed by heat treatment (39,40). These coatings performed better than plasma sprayed CoCrAlY coatings during 600 hours of testing at 1100°C in a burner rig.

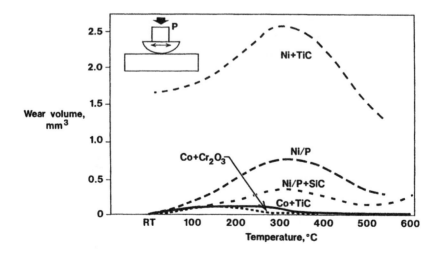

Figure 25: Wear resistance of various composite coatings as a function of temperature. Adapted from reference 38.

COMPOSITION MODULATED COATINGS

Microlayered metallic materials, sometimes referred to as composition modulated alloys, have gained general recognition as the result of their unusual and sometimes outstanding properties. These films consist of periodic repetition of thin layers of different composition with a thickness of a few nanometers and the number of such layers varying from 10 to a few 100 (41). Composition modulated alloys in a variety of binary metallic systems have been found to exhibit novel and interesting mechanical, transport, magnetic and wear properties (42). For example, the tensile strength of electrodeposited composition modulated layers of the nominal overall composition 90Ni-10Cu was shown to increase sharply to the 190,000 psi (1300 Mpa) range as the thickness of the Cu layers was decreased below about 0.4 µm. This tensile strength value is almost a factor of three greater than that measured for nickel itself, and more than a factor of two greater than the handbook value for Monel 400 (43).

Wear studies with composition modulated films suggest that the layer microstructure of these coatings may provide internal barriers to wear damage, thus leading to increases in wear resistance. Sliding wear measurements on Ni-Cu composition modulated coatings on AISI type 52100 steel showed that alternate layers of nickel and copper with equal layer spacings of either 10 or 100 nm were more wear resistant than either pure nickel or copper deposits (44-46). Figure 26 shows that the copper coating exhibited the highest wear, approximately twice that of the nickel deposit. For loads less than 18N, both composition modulated coatings exhibited lower wear than the nickel or copper coatings. For loads less than 15N, the 10 nm Ni-Cu coatings exhibited the lowest wear.

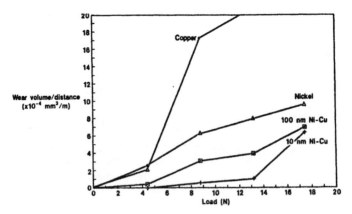

Figure 26: Wear volume per unit sliding distance for copper, nickel, and composition modulated nickel-copper deposits. Adapted from reference 44.

REFERENCES

1. M. Antler, "Wear and Contact Resistance", *Properties Of Electrodeposits; Their Measurement And Significance*, R. Sard, H. Leidheiser, Jr., and F. Ogburn, Editors, The Electrochemical Soc., 353 (1975).

2. K. G. Budinski, *Surface Engineering For Wear Resistance*, Prentice Hall (1988).

3. D. T. Gawne and U. Ma, "Wear Mechanisms in Electroless Nickel Coatings", *Wear*, 120, 125 (1987).

4. D. T. Gawne and U. Ma, "Friction and Wear of Chromium and Nickel Coatings", *Wear*, 129, 123 (1989).

5. N. Feldstein, T. Lancsek, D. Lindsay and L. Salerno, "Electroless Composite Plating", *Metal Finishing*, 81, 35 (August 1983).

6. W. F. Sharp, "Properties and Applications of Composite Diamond Coatings", *Wear*, 32, 315 (1975).

7. D. T. Gawne and T. G. P. Gudyanga, "Wear Behavior of Chromium Electrodeposits" in *Coatings And Surface Treatment For Corrosion And Wear Resistance*, K. N. Strafford, P. K. Datta and C. G. Googan, Editors, Inst. of Corrosion Science and Technology, Chapter 2 (1984).

8. K. G. Budinski, "Wear Characteristics of Industrial Platings", *Selection And Use Of Wear Tests For Coatings*, ASTM STP 769, R. G. Bayer, Editor, *American Society for Testing and Materials*, 118 (1982).

9. D. T. Gawne and N. J. Despres, "The Influence of Process Conditions on the Friction and Wear of Electrodeposited Chromium Coatings", *J. Vac. Sci. Technol.*, A3 (6), 2334 (Nov/Dec 1985).

10. R. B. Alexander, "Combined Hard Chrome Plating & Ion Implantation for Improving Tool Life", *Plating & Surface Finishing*, 77, 18 (Oct 1990).

11. W. C. Oliver, R. Hutchings and J. B. Pethica, "The Wear Behavior of Nitrogen-Implanted Metals", *Metallurgical Transactions A*, 15A, 2221 (1984).

12.   R. Hutchings, "The Subsurface Microstructure of Nitrogen-Implanted Metals", *Materials Science and Engineering*, 69, 129 (1985).

13.   K. Terashima, T. Minegishi, M. Iwaki and K. Kawashima, "Surface Modification of Electrodeposited Chromium Films by Ion Implantation", *Materials Science and Engineering*, 90, 229 (1987).

14.   W. Lohmann and J. G. P. Van Valkenhoef, "Improvement in Friction and Wear of Hard Chromium Layers by Ion Implantation", *Materials Science and Engineering*, A116, 177 (1989).

15.   R. B. Alexander, "Combined Hard Chrome Plating and Ion Implantation for Improving Tool Life", *Proceedings SUR/FIN 90*, American Electroplaters & Surface Finishers Soc., 847 (1990).

16.   B. Jackson, R. Macary and G. Shawhan, "Low Phosphorus Electroless Nickel Coating Technology", *Trans. Inst. Metal Finishing*, 68, 75 (1990).

17.   D. T. Gawne and U. Ma, "Structure and Wear of Electroless Nickel Coatings", *Mater. Sci. Technol.*, 3, 228 (March 1987).

18.   H. Wapler, T. A. Spooner and A. M. Balfour, "Diamond Coatings for Increased Wear Resistance", *Ind. Diamond Review*, 40, 251 (1979).

19.   D. D. Roshon, "Electroplated Diamond Composites for Abrasive Wear Resistance", *IBM J. of Res. Dev.*, 22, 681 (1978).

20.   M. Stevenson, Jr., "Electroless Nickel: No Longer Just a Coating", *Proceedings SUR/FIN 90*, American Electroplaters & Surface Finishers Soc., 1273 (1990).

21.   P. R. Ebdon, "The Performance of Electroless Nickel/PTFE Composites", *Plating & Surface Finishing*, 75, 65 (Sept 1988).

22.   D. A. Brockman, "Hard Chrome Plated Electroless Nickel?", *Products Finishing*, 46, 46 (Jan 1982).

23.   J. E. McCaskie, M. McNeil and A. Neiderer, "Properties of Electroless Nickel/High-Efficiency Chromium Deposits", Proceedings Electroless Nickel 89, *Products Finishing*, Cincinnati, Ohio (1989).

24.   M. Antler, "Wear, Friction, and Lubrication", *Gold Plating Technology*, F. H. Reid and W. Goldie, Editors, Electrochemical Publications, Ltd., Chapter 21 (1974).

25.   M. Antler, "What Do Gold Plating Specs Really Mean?" *Products Finishing*, 33, 56 (Oct 1969).

26.   M. Antler, "Sliding Wear of Metallic Contacts", *IEEE Trans. on Components, Hybrids and Mfg. Technology*, Vol. CHMT-4, (1), 15 (March 1981).

27.   E. J. Kudrak, J. A. Abys, I. Kadija and J. J. Maisano, "Wear Reliability of Gold-Flashed Palladium vs. Hard Gold on a High-Speed Digital Connector System", *Plating & Surface Finishing*, 78, 57 (March 1991).

28.   T. Sato, Y. Matsui, M. Okada, K. Murakawa and Z. Henmi, "Palladium With a Thin Gold Layer as a Sliding Contact Material", *IEEE Trans. on Components, Hybrids and Mfg. Technology*, Vol, CHMT-4, (1), 10 (March 1981).

29.   H. C. Angus, "Properties and Behavior of Precious Metal Electrodeposits for Electrical Contacts", *Trans. Inst. Metal Finishing*, 39, 20 (1962).

30.   F. H. Reid, "Platinum Metal Plating-A Process and Applicational Survey", *Trans. Inst. Metal Finishing*, 48, 115 (1970).

31.   L. F. Spencer, "Anodizing of Aluminum Alloys-Hardcoating", *Metal Finishing*, 66, 58 (Nov 1968).

32.   J. B. Mohler, "Abrasion Resistance and Density of Anodic Coatings", *Metal Finishing*, 69, 53 (June 1971).

33.   *The Surface Treatment And Finishing Of Aluminum And Its Alloys*, Volumes I and II, S. Wernick, R. Pinner, and P. G. Sheasby, Editors, ASM International, Metals Park, Ohio (1987).

34.   I. Machlin and N. J. Whitney, "Anodizing of Aluminum Alloys", *Metal Finishing*, 59, 55 (Feb 1961).

35.   D. J. George and J. H. Powers, "Hard Anodic Coatings: Characteristics, Applications, and Some Recent Studies on Processing and Testing", *Plating*, 56, 1240 (1969).

36.   H. G. Arlt, "The Abrasion Resistance of Anodically Oxidized Coatings on Aluminum", *ASTM Proceedings*, 40, 967 (1940).

37.   E. C. Kedward and K. W. Wright, "The Wear Control of Aircraft Parts Using A Composite Electroplate", *Plating & Surface Finishing*, 65, 38 (August 1978).

38.   M. Thoma, "A Cobalt/Chromic Oxide Composite Coating for High Temperature Wear Resistance", *Plating & Surface Finishing*, 71, 51 (Sept 1984).

39.   F. J. Honey, V. Wride and E. C. Kedward, "Electrodeposits for High Temperature Corrosion Resistance, *Plating & Surface Finishing*, 73, 42 (Oct 1986).

40.   F. J. Honey, E. C. Kedward and V. Wride, "The Development of Electrodeposits for High Temperature Oxidation/Corrosion Resistance", *J. Vac. Sci. Technol.*, A4 (6), 2593 (Nov/Dec 1986).

41.   J. P. Celis, J. R. Roos, B. Blanpain and M. Gilles, "Pulse Electrodeposition of Compositionally Modulated Multilayers", *Proc. 12th World Congress on Surface Finishing*, Paris, France, 435 (Oct 1988).

42.   J. Yahalom, D. F. Tessier, R. S. Timsit, A. M. Rosenfeld, D. F. Mitchell and P. T. Robinson, "Structure of Composition Modulated Cu/Ni Thin Films Prepared by Electrodeposition", *J. Mater. Res.*, 4, 755 (July/Aug 1989).

43.   D. Tench and J. White, "Enhanced Tensile Strength for Electrodeposited Nickel-Copper Multilayer Composites", *Metallurgical Transactions A*, 15A, 2039 (1984).

44.   A. W. Ruff, N. K. Myshkin, and Z. X. Wang, "Wear of Composition Modulated Nickel-Copper Alloys", *Proc. International Conference on Engineered Materials for Advanced Friction and Wear Applications*, Gaithersburg, MD, ASM International (March 1988).

45.   A. W. Ruff and N. K. Myshkin, "Lubricated Wear Behavior of Composition Modulated Nickel-Copper Coatings", *Journal of Tribology*, 111, 156 (Jan 1989).

46.   A. W. Ruff and Z. X. Wang, "Sliding Wear Studies of Ni-Cu Composition Modulated Coatings on Steel", *Wear*, 131, 259 (1989).

INDEX

Printed and bound by CPI Group (UK) Ltd, Croydon, CR0 4YY

03/10/2024

01040434-0008